Mathematical Theorems - Boundary Value Problems and Approximations

Edited by Lyudmila Alexeyeva

Published in London, United Kingdom

IntechOpen

Supporting open minds since 2005

Mathematical Theorems - Boundary Value Problems and Approximations
http://dx.doi.org/10.5772/intechopen.83329
Edited by Lyudmila Alexeyeva

Contributors
Gulmira K. Zakiryanova, Lyudmila Alexeyevna Alexeyeva, Yergali Kurmanov, José Alfredo Alfredo Sánchez De León, Maria Vladimirovna Shubina, SongIl Ri, Vasileios Drakopoulos, Bayegizova Aigulim, Dadayeva Assiyat, Anatoly Shatrov, Igor Andrianov

First published in London, United Kingdom, 2020 by IntechOpen
IntechOpen is the global imprint of INTECHOPEN LIMITED, registered in England and Wales, registration number: 11086078, 5 Princes Gate Court, London, SW7 2QJ, United Kingdom
Printed in Croatia

British Library Cataloguing-in-Publication Data
A catalogue record for this book is available from the British Library

Additional hard and PDF copies can be obtained from orders@intechopen.com

Mathematical Theorems - Boundary Value Problems and Approximations
Edited by Lyudmila Alexeyeva
p. cm.
Print ISBN 978-1-83880-071-0
Online ISBN 978-1-83880-072-7
eBook (PDF) ISBN 978-1-83880-341-4

For EU product safety concerns:
IN TECH d.o.o., Prolaz Marije Krucifikse Kozulić 3, 51000 Rijeka, Croatia, info@intechopen.com or visit our website at intechopen.com.

Meet the editor

Lyudmila Alexeyeva (born in 1947, Kaliningrad, Russia), Doctor of physical and mathematical sciences, professor, academician of International Eurasian Academy of Sciences, graduated with honors from the Mechanics and Mathematics Faculty of M.V. Lomonosov Moscow State University (Russia). Then she worked at the Institute of Mathematics and Mechanics of Academy of Sciences (Kazakhstan, Alma-Ata) in the laboratory of theory of seismic resistance of underground structures (1973-1991). From 1992 to the present day, she has run the Wave Dynamics Laboratory of Institute of Mathematics and Mathematical Modeling. From 2011 to 2018, she worked as the Head of Department of Mathematical Physics and Modeling in this institute. Concurrently, she worked as a Professor at al-Farabi Kazakh National University at Mechanics and Mathematics Faculty She has authored 5 monographs and over 300 scientific publications. Five doctoral and 14 candidate dissertations were defended under her supervision. For her active scientific work, she was awarded the State Scholarship for outstanding contribution to development of science and technology (four times), and a medal for labor valor.

Contents

Preface **XI**

Chapter 1 **1**
Exact Traveling Wave Solutions of One-Dimensional Parabolic-Parabolic Models of Chemotaxis
by Maria Vladimirovna Shubina

Chapter 2 **23**
Generalized and Fundamental Solutions of Motion Equations of Two-Component Biot's Medium
by Lyudmila Alexeyeva and Yergali Kurmanov

Chapter 3 **37**
Boundary Integral Equations of no Stationary Boundary Value Problems for the Klein-Gordon Equation
by Bayegizova Aigulim and Dadayeva Assiyat

Chapter 4 **57**
Singular Boundary Integral Equations of Boundary Value Problems for Hyperbolic Equations of Mathematical Physics
by Lyudmila A. Alexeyeva and Gulmira K. Zakiryanova

Chapter 5 **77**
Padé Approximation to Solve the Problems of Aerodynamics and Heat Transfer in the Boundary Layer
by Igor Andrianov and Anatoly Shatrov

Chapter 6 **99**
Alternative Representation for Binomials and Multinomies and Coefficient Calculation
by José Alfredo Sánchez de León

Chapter 7 **117**
How Are Fractal Interpolation Functions Related to Several Contractions?
by SongIl Ri and Vasileios Drakopoulos

Preface

Mathematical modeling is the most effective method for studying a wide variety of processes: physical, biological, social, and many others. The mathematical model of any process includes a set of defining parameters and characteristics of the object under study, as well as the establishment of mathematical relationships between them.

Dynamic processes in various media and structures under the action of external and internal sources of perturbations are described, as a rule, by differential equations of various types, the solutions of which depend on the geometry of the object under study and conditions on its boundaries, which can be infinite. Mathematical models of such processes are boundary value problems of mathematical physics and mechanics.

The main content of this book is related to construction of analytical solutions of differential equations and systems of mathematical physics, to the development of analytical methods for solving boundary value problems for such equations, and the study of properties of their solutions. A wide class of equations (elliptic, parabolic, and hyperbolic) is considered here, on the basis of which complex wave processes in biological and physical media can be simulated.

Chapter 1 is devoted to construction and research of solutions to a complex multiparameter system of nonlinear partial differential equations of the parabolic type and their modifications with an application to the problems of hemotaxis process in living organisms. Transport solutions of these equations are constructed that describe traveling waves of the solitons type. For various partial values of the parameters, the exact solutions of these equations are constructed using the theory of Bessel and hypergeometric functions. The constructed solutions are well illustrated by the presented graphic material.

In Chapter 2, a two-component Biot medium is considered, which allows modeling the dynamics of liquid and gas saturated porous media and rods. By using Fourier transformation of generalized functions, Green tensors of these hyperbolic systems are constructed in spaces of different dimensions. The cases of nonstationary motion and periodic vibration are considered. The regular integral representations of these solutions are given for acting regular and singular mass forces.

Chapter 3 is devoted to solving the boundary value problems for equations of hyperbolic type of theoretical physics, which describes the motion of elementary particles in potential fields. In particular, the Klein-Gordon equation is considered, whose solutions for various scalar fields have been studied by many authors.

Note that boundary value problems for hyperbolic equations and systems in domains with arbitrary boundary geometry are among the most complex problems of mathematical physics, since the classical potential theory, characteristic of solving boundary value problems for elliptic and parabolic systems, is not applicable in the initial space-time. This is due to singularity and hyper singularity of the fundamental solutions of hyperbolic equations on wave fronts, as well as their belonging to the class of singular generalized functions in spaces with odd dimensions, which makes it impossible to apply classical methods of potential theory for such BVPs.

The method of generalized functions, used in Chapter 3 for solving the boundary value problems, allows construction of regular integral representations of solutions, which determine the solution inside the domain through the boundary values of the solution and their derivatives. Some of them are known from boundary conditions, and to determine the unknown boundary functions, the resolving singular boundary integral equations are constructed in 2D and 3D spaces. The correctness of the posed problems is proved, taking into account the appearance of shock waves.

Chapter 4 is devoted to the development of the method of generalized functions (GFM) for construction of solutions of BVPs for hyperbolic systems of mathematical physics, which describe wave processes and dynamics of continue media, in particular, dynamics of elastic solids and media.

This very constructive method is based on the idea of transition from the classical formulation of initial boundary value problems to its formulation in the space of generalized functions. This allows reducing a process of BVP solving to solving the differential equations system with a singular right-hand side in the space of generalized functions. This singular part contains simple and double layers, the densities of which are determined by the value of the solution and its derivatives on the boundary. By using the Green matrix (tensor) of these equations, a generalized solution of the BVP can be obtained in the form of a convolution of the right-hand side with this matrix. The regularization of solutions and transition to their regular representations makes it possible to construct a classical solution for the BVP. The asymptotic properties of Green tensor and the tensors derived from it, which are the kernels of these integral representations, make it possible to construct resolving singular boundary integral equations.

The method of generalized functions is universal and can be applied to differential equations of any types. It allows us to study processes accompanied by shock waves, which is often impossible using classical methods. All these issues are considered in sufficient detail in this chapter.

The next part of this book covers the construction of various approximations for solutions of differential equations and functions. Note that the use of various approximations in the form of series and sequences of elementary and special functions to construct solutions to equations and boundary value problems is one of the most common ways to solve them, and the choice of such approximations is closely related to the specifics of the problems being solved.

The Padé approximation method is one of the most promising nonlinear methods for summing power series and localizing its singular points. It is convenient to use it when constructing solutions of equations based on asymptotic expansions of solutions in a small parameter in the vicinity of singular points.

Chapter 5 is devoted to applications of asymptotic methods in solving the nonlinear BVPs of mechanics using the Padé approximation. Here three boundary value problems for nonlinear ordinary differential equations are considered: the Airy boundary value problem, the BVP for Blasius equation, which describes laminar flow of boundary layers, and BVP for equations of laminar boundary layer near a semi-infinite plate in super-sonic flow of viscous perfect gas. In this chapter, asymptotic interior and exterior Padé approximations for these problems are constructed. In the last problem, the influence of the Mach number on asymptotic of the solutions is also investigated.

Chapter 6 is expository for the importance of using programming in algebraic calculations. Although complete binomial and multinomial construction can be a hard task, there exist some mathematical formulas that can be deployed to calculate binomial and multinomial coefficients, in order to make it quicker. A main aim here is the development of an alternative method to carry out the calculation of binomial and multinomial coefficients.

The analytic formulas are presented, that yield binomial coefficients, by means of summation series, and the equation targeted at binomial calculations, is deduced which is convenient for calculations. Finally an algorithm set up on Computing Algebra System (CAS) Maxima is raised. The Appendix explains all calculus and logic deductions in this chapter. This algorithm checks if it is faster than usual one by calculations.

Chapter 7 is devoted to problems of interpolation of nondifferential functions, which are typical for real and experimental signals by study of different physical processes and others. The authors provide an overview of several types of fractal interpolation functions. The connections between fractal interpolation functions resulting from Banach contractions as well as those resulting from Rakotch contractions are considered. The theoretical and practical significance for the generation of fractal functions for interpolation purposes in 2D and 3D spaces is given. The new methods presented can be extended to piecewise fractal interpolation functions.

The book would be interesting for specialists in the field of mathematical and theoretical physics, mechanics and biophysics, students of mechanics, mathematics, physics and biology departments of higher educational institutions. The thoughtful reader will find in it a lot that is necessary and useful for his scientific research work.

Lyudmila Alexeyeva
Institute of Mathematics and Mathematical Modeling,
Kazakhstan

Chapter 1

Exact Traveling Wave Solutions of One-Dimensional Parabolic-Parabolic Models of Chemotaxis

Maria Vladimirovna Shubina

Abstract

In this chapter we consider several different parabolic-parabolic systems of chemotaxis which depend on time and one space coordinate. For these systems we obtain the exact analytical solutions in terms of traveling wave variables. Not all of these solutions are acceptable for biological interpretation, but there are solutions that require detailed analysis. We find this interesting, since chemotaxis is present in the continuous mathematical models of cancer growth and invasion (Anderson, Chaplain, Lolas, et al.) which are described by the systems of reaction–diffusion-taxis partial differential equations, and the obtaining of exact solutions to these systems seems to be a very interesting task, and a more detailed analysis is possible in a future study.

Keywords: parabolic-parabolic system, exact solution, soliton solution, Patlak-Keller-Segel model, chemotaxis

1. Introduction

This chapter uses the publications of Shubina M.V.:

1. Exact Traveling Wave Solutions of One-Dimensional Parabolic-Parabolic Models of Chemotaxis, Russian J Math Phys., Maik Nauka/Interperiodica Publishing (Russian Federation), 25(3), 383–395, 2018.

2. The 1D parabolic-parabolic Patlak-Keller-Segel model of chemotaxis: The particular integrable case and soliton solution, J Math Phys., 57(9), 091501, 2016.

Chemotaxis, or the directed cell (bacteria or other organisms) movement up or down a chemical concentration gradient, plays an important role in many biological and medical fields such as embryogenesis, immunology, cancer growth, and invasion. The macroscopic classical model of chemotaxis was proposed by Patlak in 1953 [1] and by Keller and Segel in the 1970s [2–4]. Since then, the mathematical modeling of chemotaxis has been widely developed. This model is described by the system of coupled nonlinear partial differential equations. Proceeding from the study of the properties of these equations, it is concluded that the model demonstrates a deep mathematical structure. The survey of Horstmann [5] provides a detailed introduction into the mathematics of the Patlak-Keller-Segel model and summarizes different mathematical results; the detailed reviews also can be found in the textbooks of Suzuki [6] and Perthame [7]. In the review of Hillen and Painter [8], a number of variations of the original Patlak-Keller-Segel model are explored in detail. The authors study their formulation from a biological perspective, summarize key results on their analytical properties, and classify their solution forms [8]. It should be noted that interest in the Patlak-Keller-Segel model does not weaken and new works appear devoted to the study of various properties of equations and their solutions [9–12] and the links below.

In this chapter we investigate a number of different models describing chemotaxis. The aim of this paper is to obtain exact analytical solutions of these models. For one-dimensional parabolic-parabolic systems under consideration, we present these solutions in explicit form in terms of traveling wave variables. Of course, not all of the solutions obtained can have appropriate biological interpretation since the biological functions must be nonnegative in all domains of definition. However some of these solutions are positive and bounded, and their analysis requires further investigation. Despite the large number of works devoted to the systems under consideration and their properties, as well as the properties of their solutions, it seems to us that the solutions obtained in this paper are new.

The Patlak-Keller-Segel model describes the space–time evolution of a cell density $u(t, \vec{r})$ and a concentration of a chemical substance $v(t, \vec{r})$. The general form of this model is:

$$\begin{cases} u_t - \nabla(\delta_1 \nabla u - \eta_1 u \nabla \phi(v)) = 0 \\ v_t - \delta_2 \nabla^2 v - f(u, v) = 0, \end{cases}$$

where $\delta_1 > 0$ and $\delta_2 \geq 0$ are cell and chemical substance diffusion coefficients, respectively, and η_1 is a chemotaxis coefficient; when $\eta_1 > 0$, this is an attractive chemotaxis ("positive taxis"), and when $\eta_1 < 0$, this is a repulsive ("negative") one [13, 14]. $\phi(v)$ is the chemosensitivity function, and $f(u, v)$ characterizes the chemical growth and degradation. These functions are taken in different forms that correspond to some variations of the original Patlak-Keller-Segel model. We follow the reviews of Hillen and Painter [8] and of Wang [15] and consider the models presented therein.

This paper is concerned with one-dimensional simplified models when the coefficients δ_1, δ_2, and η_1 are positive constants, $x \in \Re, t \geq 0$, $u = u(x, t)$, and $v = v(x, t)$.

2. Signal-dependent sensitivity model

Let us start with a model that allows nonnegative bounded solutions that may be of interest from a biological point of view. Now consider the "logistic" model, one of versions of signal-dependent sensitivity model [8] with the chemosensitivity functions $\phi(v) = (1 + b) \ln(v + b)$, where $b = const$, and $f(u, v) = \tilde{\sigma} u - \tilde{\beta} v$. In the review [5] one can see a mathematical analysis of this model. When $b = 0$ and $\tilde{\beta} = 0$, the existence of traveling waves was established in [16, 17]. The replacements of $t \to \delta_1 t$ and $u \to \sigma \frac{\tilde{\sigma}}{\delta_1} u$ give $\delta_1 = 1$, $\alpha = \frac{\delta_2}{\delta_1}$, $\beta = \frac{\tilde{\beta}}{\delta_1}$, and $\sigma = \pm 1$. We also set $\eta = \frac{\eta_1(1+b)}{\delta_1}$, $1 + b > 0$, as well as $\phi(v) = \ln|v + b|$. It should be noted that a sign of σ may effect on the mathematical properties of the system. So, $\sigma = 1$ corresponds to an increase of a chemical substance, proportional to cell density, whereas $\sigma = -1$ corresponds to its decrease. And as we shall see later, various solutions correspond to these two cases.

After the above replacements, the model reads:

$$\begin{cases} u_t - u_{xx} + \eta\left(u\frac{v_x}{v+b}\right)_x = 0 \\ v_t - \alpha v_{xx} - \sigma u + \beta v = 0. \end{cases} \tag{1}$$

If we introduce the function $\upsilon = v + b$, in terms of traveling wave variable $y = x - ct$, where $c = const$, this system has the form:

$$\begin{cases} u_y + cu - \eta u(\ln(\upsilon))_y + \lambda = 0 \\ \alpha\upsilon_{yy} + c\upsilon_y - \beta\upsilon + \beta b + \sigma u = 0, \end{cases} \tag{2}$$

where $u = u(y)$, $\upsilon = \upsilon(y)$, and λ is an integration constant.
In this chapter we will consider the case of $\lambda = 0$. Then Eq. (2) gives:

$$u = C_u e^{-cy}\upsilon^\eta, \tag{3}$$

C_u is a constant and we will examine the following equation for υ:

$$\alpha\upsilon_{yy} + c\upsilon_y - \beta\upsilon + \beta b + \sigma C_u e^{-cy}\upsilon^\eta = 0. \tag{4}$$

Since η is a positive constant, we consider two cases: $\eta = 1$ [Eq. (4) is a linear nonhomogeneous equation] and $\eta \neq 1$.

A. $\eta = 1$

Let us begin with $\eta = 1$. We introduce the new variable z and the new function w:

$$\begin{aligned} z &= \left(\frac{4\sigma C_u}{\alpha c^2}\right)^{\frac{1}{2}} e^{-\frac{cy}{2}} \\ w &= \left(\frac{4\sigma C_u}{\alpha c^2}\right)^{\frac{\alpha-2}{4\alpha}} \upsilon e^{\frac{cy}{2\alpha}} \end{aligned} \tag{5}$$

and Eq. (4) becomes:

$$z^2 w_{zz} + z w_z + w\left(z^2 - \nu^2\right) = \Lambda z^{-\frac{1}{\alpha}}, \tag{6}$$

where $\nu^2 = \frac{1}{\alpha^2}\left(1 + \frac{4\alpha\beta}{c^2}\right)$ and $\Lambda = -\frac{4\beta b}{\alpha c^2}\left(\frac{4\sigma C_u}{\alpha c^2}\right)^{\frac{1}{4}}$. Eq. (6) is the Lommel differential equation [18, 19] with $\mu = -1 - \frac{1}{\alpha}$, and we consider $\sigma C_u > 0$. Since this is a linear inhomogeneous second-order differential equation, one can integrate it by the method of variation of parameters. We assume a solution in the form:

$$w(z) = C_J(z)J_\nu(z) + C_Y(z)Y_\nu(z),$$

where $J_\nu(z)$ and $Y_\nu(z)$ are Bessel functions and $C_J(z)$ and $C_Y(z)$ are the functions of z that satisfy the equations:

$$\begin{aligned} J_\nu(z)\left(C_J(z)\right)_z + Y_\nu(z)\left(C_Y(z)\right)_z &= 0 \\ \left(J_\nu(z)\right)_z\left(C_J(z)\right)_z + \left(Y_\nu(z)\right)_z\left(C_Y(z)\right)_z &= \Lambda z^{-\frac{1}{\alpha}}. \end{aligned}$$

Considering that Wronskian $W(J_\nu, Y_\nu)(z) = \frac{2}{\pi z}$, we obtain:

$$C_J(z) = c_J - \frac{\Lambda\pi}{2}\int z^{-1-\frac{1}{\alpha}} Y_\nu(z)dz$$

$$C_Y(z) = c_Y + \frac{\Lambda\pi}{2}\int z^{-1-\frac{1}{\alpha}} J_\nu(z)dz,$$

where c_J and c_Y are constants. If both of the numbers $-\frac{1}{\alpha} \pm \nu$ are positive, the lower limits in the integrals may be taken to be zero. Then a particular integral of Lommel equation "proceeding in ascending powers of z" is $s_{\mu,\nu}(z)$ [19]; if one considers a solution of Lommel equation "in the form of descending series," one obtains the function $S_{\mu,\nu}(z)$ [19] [see Eq. (8)]. Thus, quoting Watson [19] "...and so, of Lommel's two functions $s_{\mu,\nu}(z)$ and $S_{\mu,\nu}(z)$, it is frequently more convenient to use the latter." Then the general solution of Eq. (6) has the form:

$$w(z) = C_J J_\nu(z) + C_Y Y_\nu(z) + \Lambda S_{\mu,\nu}(z), \tag{7}$$

where C_J and C_Y are constants,

$$S_{\mu,\nu}(z) = s_{\mu,\nu}(z) + 2^{\mu-1}\Gamma\left(\frac{\mu-\nu+1}{2}\right)\Gamma\left(\frac{\mu+\nu+1}{2}\right)$$
$$\left[\sin\left(\frac{\pi}{2}(\mu-\nu)\right)J_\nu(z) - \cos\left(\frac{\pi}{2}(\mu-\nu)\right)Y_\nu(z)\right], \tag{8}$$
$$s_{\mu,\nu}(z) = \frac{z^{\mu+1}}{\left[(\mu+1)^2-\nu^2\right]}\,{}_1F_2\left(1;\frac{\mu-\nu+3}{2},\frac{\mu+\nu+3}{2};-\frac{z^2}{4}\right)$$

are Lommel functions, and ${}_1F_2$ is the generalized hypergeometric function [18, 19]. Further, substituting the initial variable y and the function v [see Eq. (5)] into Eq. (7), we obtain a formal solution.

1. $b = 0$

We first consider the case $b = 0$. Then $\upsilon = \nu \geq 0$ and $C_u > 0$. Eq. (6) becomes homogeneous, and for $\sigma = 1$, its general solution is:

$$w(z) = C_J J_\nu(z) + C_Y Y_\nu(z). \tag{9}$$

However one can check that the function $u = u(y)$ diverges as $cy \to -\infty$ for all ν.

Consider now $\sigma = -1$. For $v(y)$ to be real, let $\alpha = 2$. Then Eq. (6) becomes the modified Bessel equation; the analysis of solution behavior at $\pm\infty$ leads to suitable solutions for $v(y)$ and $u(y)$:

$$v(y) = e^{-\frac{cy}{4}}K_\nu\left(\sqrt{\frac{2C_u}{c^2}}e^{-\frac{cy}{2}}\right)$$
$$u(y) = C_u e^{-\frac{5cy}{4}}K_\nu\left(\sqrt{\frac{2C_u}{c^2}}e^{-\frac{cy}{2}}\right) \tag{10}$$

with restrictions $\nu \leq \frac{1}{2}$ and $\beta \leq 0$. So one can see that $v(y) \to 0$ as $cy \to -\infty$ for all $\nu \leq \frac{1}{2}$; $v(y) \to 0$ for $\nu < \frac{1}{2}$ and $v(y) \to \sqrt[4]{\frac{\pi^2 c^2}{8C_u}}$ for $\nu = \frac{1}{2}$ as $cy \to \infty$ and $u(y) \to 0$ as $y \to \pm\infty$ for all $\nu \leq \frac{1}{2}$. The curves of these functions are presented in **Figures 1** and **2**, and

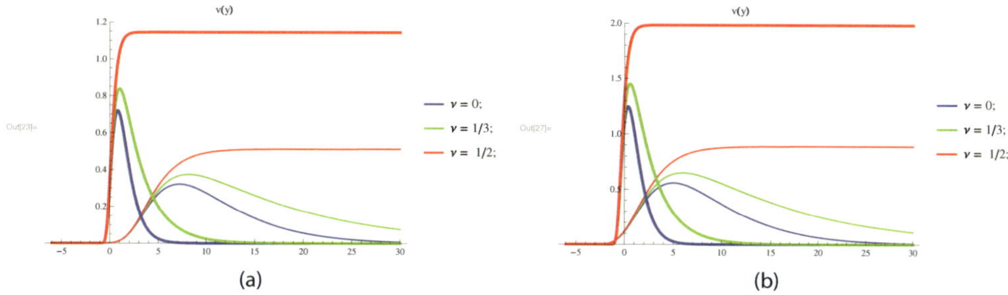

Figure 1.
(a) $v(y)$; $c = 1$; $c = 5$; $C_u = 18$. (b) $v(y)$; $c = 1$; $c = 5$; $C_u = 2$.

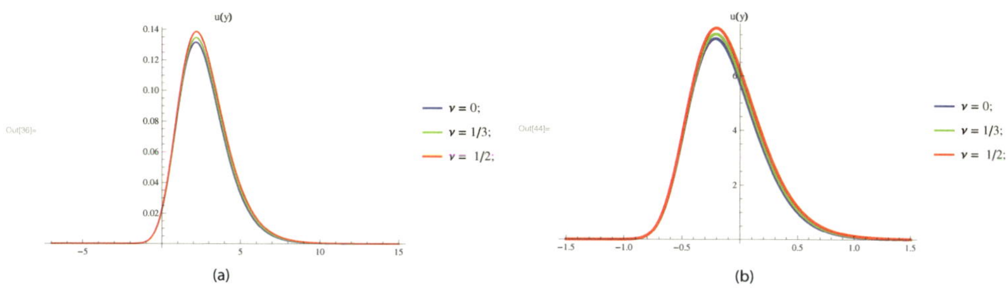

Figure 2.
(a) $u(y)$; $c = 1$; $C_u = 18$. (b) $u(y)$; $c = 5$; $C_u = 18$.

the plots for $c = 5$ are thicker than for $c = 1$. Thus, the solution obtained may be considered as a biologically appropriated one, and this requires further investigation.

2. $b > 0$

Let us return to Eq. (6) with $\Lambda \neq 0$. The analysis of solution asymptotic forms at $\pm\infty$ [18, 19] gives the following expressions for $v(y)$ and $u(y)$:

$$v(y) + b = -\frac{4\beta b}{\alpha c^2}\left(\frac{4\sigma C_u}{\alpha c^2}\right)^{\frac{1}{2\alpha}} e^{-\frac{cy}{2\alpha}} S_{\mu,\nu}\left(\sqrt{\frac{4\sigma C_u}{\alpha c^2}}\, e^{-\frac{cy}{2}}\right)$$
$$u(y) = -C_u \frac{4\beta b}{\alpha c^2}\left(\frac{4\sigma C_u}{\alpha c^2}\right)^{\frac{1}{2\alpha}} e^{-cy\left(1+\frac{1}{2\alpha}\right)} S_{\mu,\nu}\left(\sqrt{\frac{4\sigma C_u}{\alpha c^2}}\, e^{-\frac{cy}{2}}\right) \qquad (11)$$

with $\sigma C_u > 0$ and $\nu < \frac{1}{\alpha}$. The latter condition leads to the requirement $-\frac{c^2}{4\alpha} \leq \beta < 0$. The $v(y) \to -b$, $u(y) \to -\frac{\beta b}{\sigma}$ as $cy \to -\infty$ and $v(y) \to 0$ and $u(y) \to 0$ as $cy \to \infty$. Thus, one can see that for $b > 0$, $\sigma = 1$, and $C_u > 0$, $u(y) \geq 0$ is satisfied but $v(y) < 0$. These functions are presented in **Figures 3** and **4**. It should be noted that $\nu \neq \frac{1}{\alpha}$ or $\beta \neq 0$ because of the pole in Γ function.

3. $b < 0$

Using the analysis of Eq. (11), one can see that the condition $b < 0$ along with $\sigma = -1$ and $C_u < 0$ $(\sigma C_u > 0)$ leads to the fact that the function $u(y)$ has not changed, but $v(y)$ becomes positive on all domains of definition. This function is presented in **Figure 5**.

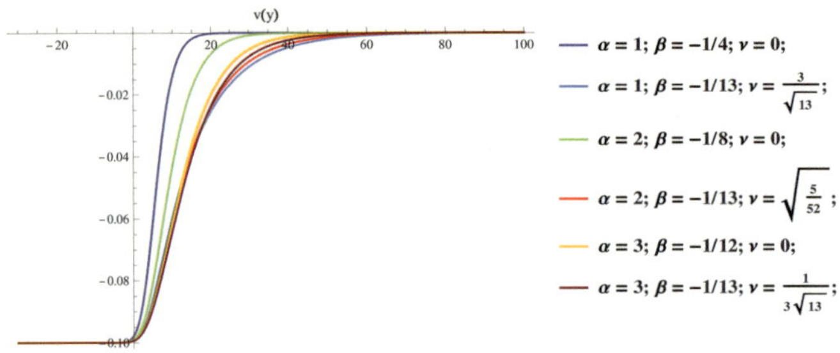

Figure 3.
$v(y)$; $c = 1$; $C_u = 9$; $\sigma = 1$; $b = 0.1$.

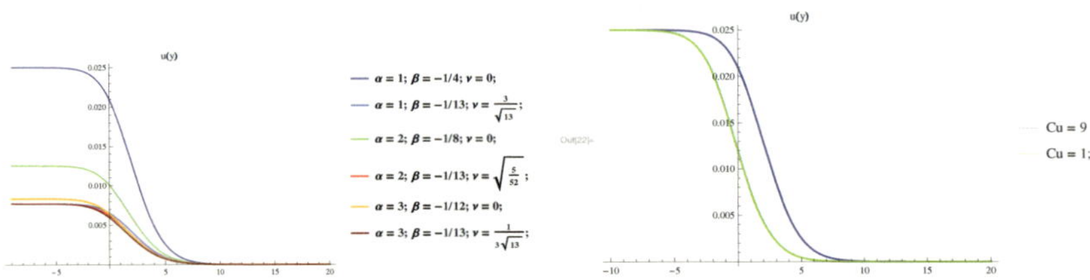

Figure 4.
(a) $u(y)$; $c = 1$; $C_u = 9$; $\sigma = 1$; $b = 0.1$. (b) $u(y)$; $c = 1$; $\sigma = 1$; $b = 0.1$; $\alpha = 1$; $\beta = -1/4$; $\nu = 0$.

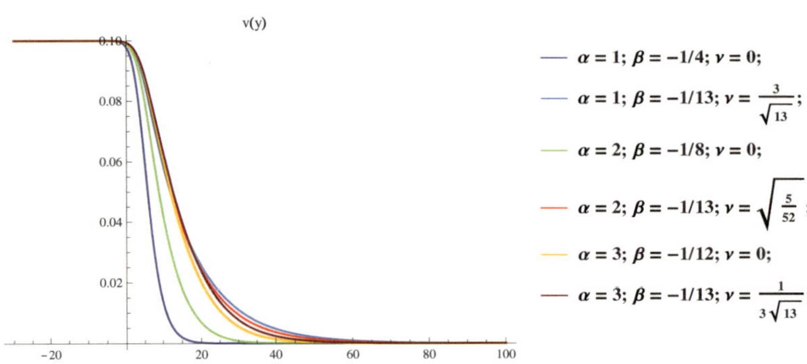

Figure 5.
$v(y)$; $c = 1$; $C_u = -9$; $\sigma = -1$; $b = -0.1$.

B. $\eta \neq 1$

Let us return to Eq. (4) and rewrite it in terms of the variable $\xi = e^{-\frac{cy}{\alpha}}$:

$$\xi^2 v_{\xi\xi} - \frac{\alpha\beta}{c^2} v + \frac{\sigma\alpha C_u}{c^2} \xi^\alpha v^\eta = -\frac{\alpha\beta b}{c^2}. \tag{12}$$

To integrate this equation, we use the Lie group method of infinitesimal transformations [20]. We find a group invariant of a second prolongation of one-parameter symmetry group vector of (12), and with its help, we transform Eq. (12) into an equation of the first order. It turns out that nontrivial symmetry group requires some conditions:

$$\begin{gathered} \frac{\alpha\beta b}{c^2} = 0, \\ \beta = \frac{(\alpha - 2)(\alpha + \eta + 1)c^2}{\alpha(\eta + 3)^2} \end{gathered} \tag{13}$$

and we consider the case $b = 0$. Thus, $v = v$, and for:

$$z = \frac{v^{\frac{1-\eta}{\alpha}}}{y}$$
$$w = v_y\, v^{-\frac{\alpha+\eta-1}{\alpha}} \tag{14}$$

we obtain the Abel equation of the second kind:

$$w_z\,[(1-\eta)w - \alpha z] + (\alpha+\eta-1)z^{-1}w^2 + \alpha z\left(-\frac{\alpha\beta}{c^2} + \frac{\sigma\alpha C_u}{c^2}z^{-\alpha}\right) = 0. \tag{15}$$

Then we find the solutions of Eq. (15) in parametric form [21] with the parameter t. Now we consider the case $2\alpha+\eta \neq 1$. A combination of substitutions leads to:

$$z = \left(-\frac{(\eta+3)\left[(\eta+1)t^2 + \frac{2\sigma\alpha C_u}{c^2}\right]}{2(2\alpha+\eta-1)}\frac{\vartheta_t(t)}{\vartheta(t)}\right)^{\frac{2}{\alpha}}$$
$$w = z^{\frac{2-\alpha}{2}}\left(t + \frac{2(2\alpha+\eta+1)}{(\eta-1)(\eta+3)}z^{\frac{\alpha}{2}}\right) + \frac{\alpha}{1-\eta}z, \tag{16}$$

where we take

$$\vartheta(t) > 0 \text{ and } (2\alpha+\eta-1)\,\vartheta_t(t) < 0, \tag{17}$$

and Eq. (15) becomes an equation for the function $\vartheta(t)$. Solving it, for $\sigma C_u > 0$, we obtain:

$$\vartheta(t) = \tilde{C}_\vartheta\left(\frac{2\sigma\alpha C_u}{c^2}\right)^{-\frac{\eta+3}{2(\eta+1)}} t\,{}_2F_1\left(\frac{1}{2}, \frac{\eta+3}{2(\eta+1)}; \frac{3}{2}; -\frac{(\eta+1)c^2}{2\sigma\alpha C_u}t^2\right) + C_\vartheta, \tag{18}$$

where $\tilde{C}_\vartheta$ and C_ϑ are constants and ${}_2F_1$ is the hypergeometric Gauss function. Further we obtain the solutions of initial Eqs. (3)–(4) in parametric form:

$$y(t) = -\frac{\alpha(\eta+3)}{c(2\alpha+\eta-1)}\ln(\vartheta(t))$$
$$v(t) = \left(-\frac{\tilde{C}_\vartheta(\eta+3)}{2(2\alpha+\eta-1)}\right)^{\frac{2}{1-\eta}}\left((\eta+1)t^2 + \frac{2\sigma\alpha C_u}{c^2}\right)^{-\frac{1}{\eta+1}}(\vartheta(t))^{\frac{2-\alpha}{2\alpha+\eta-1}} \tag{19}$$
$$u(t) = C_u\left(-\frac{\tilde{C}_\vartheta(\eta+3)}{2(2\alpha+\eta-1)}\right)^{\frac{2}{1-\eta}}\left((\eta+1)t^2 + \frac{2\sigma\alpha C_u}{c^2}\right)^{-\frac{1}{\eta+1}}(\vartheta(t))^{\frac{\alpha\eta+2\alpha+2}{2\alpha+\eta-1}}$$

where the constant $\tilde{C}_\vartheta$ is chosen so that $(2\alpha+\eta-1)\tilde{C}_\vartheta < 0$, which is consistent with Eq. (17). Using the asymptotic representation of hypergeometric Gauss function as $t \to \pm\infty$ [18], we can take:

$$C_\vartheta > |\tilde{C}_\vartheta|\,\frac{\pi}{2\sqrt{\eta+1}}\left(\frac{2\sigma\alpha C_u}{c^2}\right)^{-\frac{1}{\eta+1}}\frac{\Gamma\left(\frac{1}{\eta+1}\right)}{\Gamma\left(\frac{\eta+3}{2(\eta+1)}\right)} \tag{20}$$

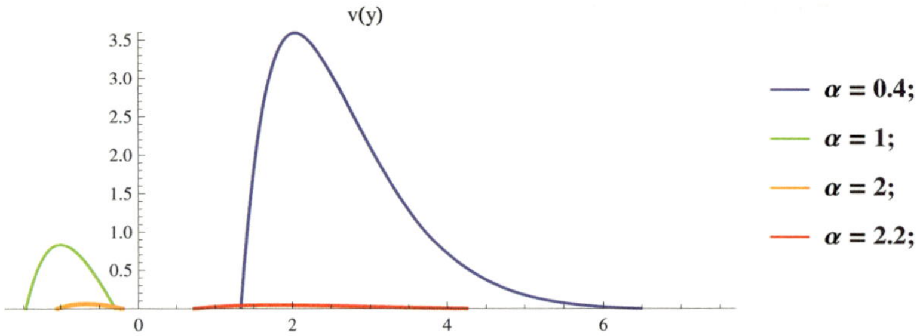

Figure 6.
$v(y)$; $\eta = 0.1$; $\frac{\sigma\alpha C_u}{c^2} = 2$; $c = 1$; $C_\vartheta = 1.4$; $|\tilde{C}_\vartheta| = 1$.

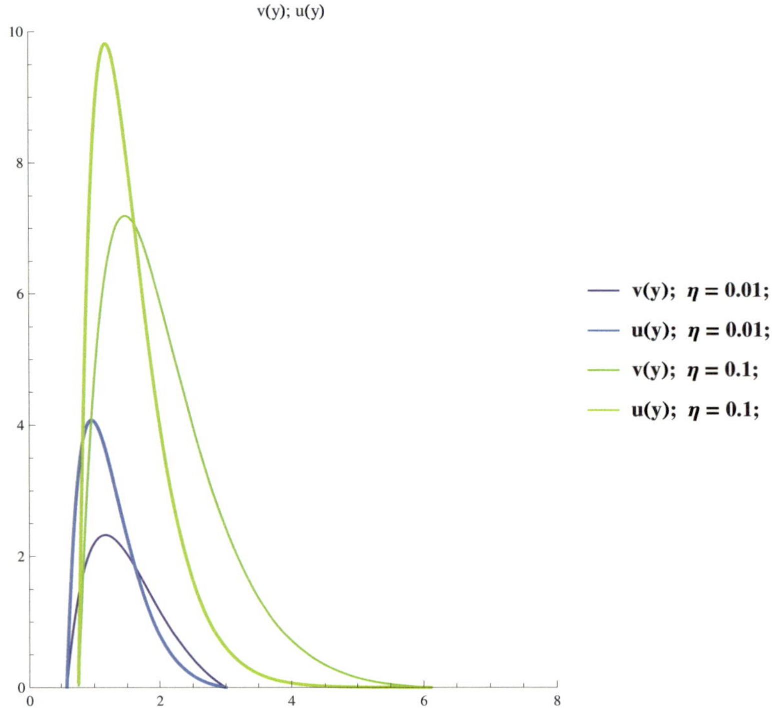

Figure 7.
$v(y)$; $u(y)$; $\alpha = 0.4$; $\frac{\sigma\alpha C_u}{c^2} = 2$; $c = 1$; $C_\vartheta = 1.35$; $|\tilde{C}_\vartheta| = 1$.

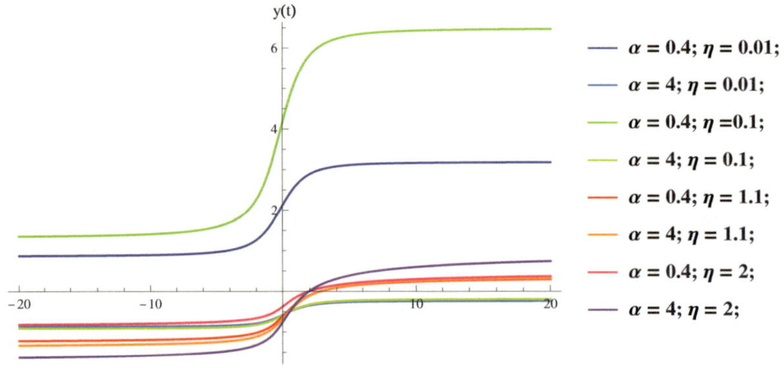

Figure 8.
$y(t)$; $\frac{\sigma\alpha C_u}{c^2} = 2$; $c = 1$; $C_\vartheta = 1.4$; $|\tilde{C}_\vartheta| = 1$.

in order for y, v, and u to be real. Then one can see that all functions in Eq. (19) are continuous bounded ones for $t \in \mathfrak{R}$ and v, u are positive. Hence, one may try to biologically interpret the functions $v(y)$ and $u(y)$, and this requires further investigation. In **Figure 6** one may see the different curves $v(y)$ for $\eta = 0.1$ and different α. **Figure 7** demonstrates $v(y)$ and $u(y)$ for two values $\eta : \eta = 0.1$ and $\eta = 0.01$, see **Figure 7**. Further, for larger values of α and η, it seems more convenient to present the curves $y(t), v(t)$, and $u(t)$ to analyze them (see **Figures 8–10**). One can see from Eq. (13) that $\beta \gtrless 0$ when $\alpha \gtrless 2$, and the case of $\beta = 0$ and $\alpha = 2$ is presented in **Figure 11**.

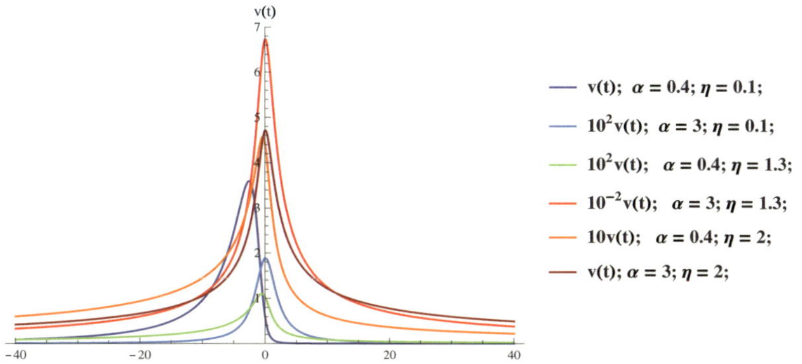

Figure 9.
$v(t)$; $\frac{\sigma\alpha C_u}{c^2} = 2$; $c = 1$; $C_\vartheta = 1.4$; $|\tilde{C}_\vartheta| = 1$.

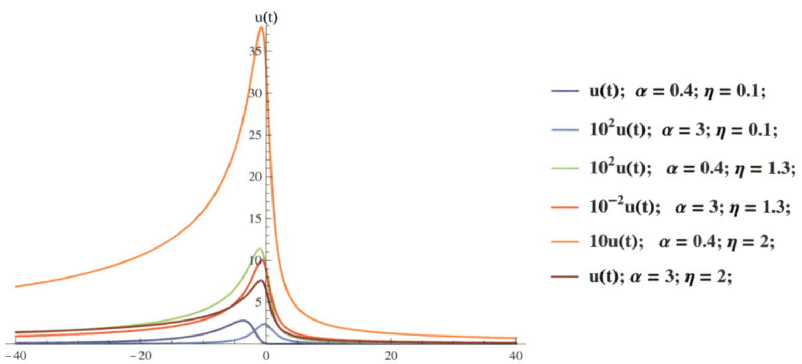

Figure 10.
$u(t)$; $\frac{\sigma\alpha C_u}{c^2} = 2$; $c = 1$; $C_\vartheta = 1.4$; $|\tilde{C}_\vartheta| = 1$.

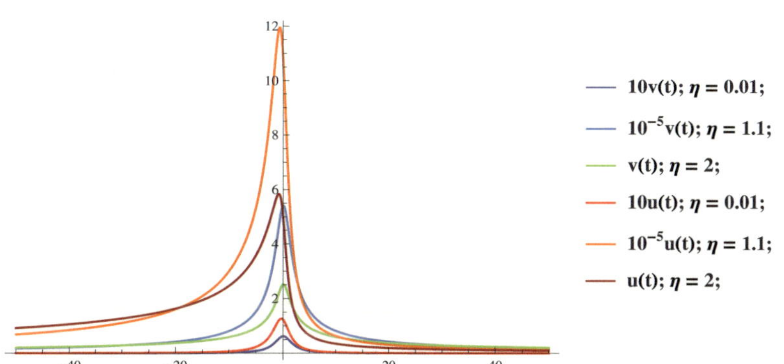

Figure 11.
$v(t)$; $u(t)$; $\alpha = 2$; $\frac{\sigma\alpha C_u}{c^2} = 2$; $c = 1$; $C_\vartheta = 1.4$; $|\tilde{C}_\vartheta| = 1$.

3. Logarithmic sensitivity

The model with logarithmic chemosensitivity function $\phi(v) \sim \ln v$ is also studied. For the case of $f(u,v) = -v^m u + \tilde{\beta} v$, where $\tilde{\beta} = const$, an extensive analysis is performed in [15]. This survey is focused on different aspects of traveling wave solutions. When $m = 0$, this model coincides with Eq. (1) for $b = 0$. When $\tilde{\beta} = 0$ and $m = 1$, the system was studied in [22, 23]. The complete analysis for $\tilde{\beta} = 0$ is performed in [15]. An existence of global solution is established in [24].

Now we consider the system with $\phi(v) = \ln v$ and $f(u,v) = \tilde{\sigma} vu - \tilde{\beta} v$. Similarly, a replacement of $t \to \delta_1 t$ and $u \to \sigma \frac{\tilde{\sigma}}{\delta_1} u$ gives $\delta_1 = 1, \eta = \frac{\eta_1}{\delta_1}, \alpha = \frac{\delta_2}{\delta_1}, \beta = \frac{\tilde{\beta}}{\delta_1}$, and $\sigma = \pm 1$. Then the model has the form:

$$\begin{cases} u_t - u_{xx} + \eta \left(u \frac{v_x}{v}\right)_x = 0 \\ v_t - \alpha v_{xx} - \sigma v u + \beta v = 0. \end{cases} \tag{21}$$

Let us rewrite the system (21) in terms of the function $\upsilon(x,t) = \ln v(x,t)$:

$$\begin{cases} u_t - u_{xx} + \eta (u \upsilon_x)_x = 0 \\ \upsilon_t - \alpha \upsilon_{xx} - \alpha (\upsilon_x)^2 + \beta - \sigma u = 0, \end{cases} \tag{22}$$

Then in terms of the traveling wave variable $y = x - ct$, where $c = const$, Eq. (22) has the form:

$$\begin{cases} u_y + cu - \eta u \upsilon_y + \lambda = 0 \\ \alpha \upsilon_{yy} + \alpha \left(\upsilon_y\right)^2 + c\upsilon_y - \beta + \sigma u = 0, \end{cases} \tag{23}$$

where $u = u(y)$, $\upsilon = \upsilon(y)$, and λ is an integration constant. To integrate Eq. (23), we tested this system on the Painlevé ODE test. One can show that for $\eta > 0$, it passes this test only if $\alpha = 2$ with the additional condition $\lambda = -\sigma c \beta \left(1 + \frac{\eta}{2}\right)$ [25]. If we express $u(y)$ as $\upsilon(y)$ from Eq. (23), we obtain an equation only for $\upsilon(y)$; for $\alpha = 2$, it has the form:

$$2\upsilon_{yyy} + 3c\upsilon_{yy} + \left(c^2 + \eta\beta\right)\upsilon_y + 2(2 - \eta)\upsilon_y \upsilon_{yy} + 2(2 - \eta)\left(\upsilon_y\right)^2 - 2\eta\left(\upsilon_y\right)^3 - c\beta - \sigma\lambda = 0. \tag{24}$$

For $\lambda = -\sigma c \beta \left(1 + \frac{\eta}{2}\right)$, this equation can be linearized. It becomes equivalent to the following linear equation for F:

$$F_y + cF = 0, \text{ where } F(y) = e^{2\upsilon}\left(2\upsilon_{yy} + c\upsilon_y - \eta\left(\upsilon_y\right)^2 + \frac{\eta\beta}{2}\right) \tag{25}$$

that gives the equation for $\upsilon(y)$:

$$2\upsilon_{yy} + c\upsilon_y - \eta\left(\upsilon_y\right)^2 + \frac{\eta\beta}{2} = C_F e^{-2\upsilon - cy}, \tag{26}$$

where $C_F = const$. If we rewrite Eq. (26) in terms of the variable $\xi = e^{-\frac{cy}{2}}$ for the function $\Psi(\xi) = e^{-\frac{\eta}{2}\upsilon}$, we obtain an equation similar to Eq. (12) with zero right-hand side:

$$\xi^2\Psi_{\xi\xi} - \frac{\eta^2\beta}{2c^2}\Psi + \frac{\eta C_F}{c^2}\xi^2\Psi^{\frac{4}{\eta}+1} = 0. \tag{27}$$

Using the result of the symmetry group analysis of Eq. (12), we can write the solution for $\beta = 0$ [see Eq. (19)]:

$$\begin{aligned} y(t) &= -\frac{2}{c}\ln(\vartheta(t)) \\ v(t) &= \frac{|\tilde{C}_\vartheta|}{2}\left(\frac{2(\eta+2)}{\eta}t^2 + \frac{2\eta C_F}{c^2}\right)^{\frac{1}{\eta+2}} \end{aligned} \tag{28}$$

where $\vartheta(t)$ is given in Eq. (18) and $u(y)$ may be expressed from Eq. (23). However one may see that $v \to \infty$ as $t \to \pm\infty$, and this solution is unacceptable as a biological function.

Another possibility to solve this equation exactly is to put C_F equal to zero. When $C_F = 0$, that means $F(y) = 0$, for $\beta \neq 0$; Eq. (27) can be linearized by $\xi = e^\tau$ [21]. Its solution has three forms according to a sign of the expression $D = \frac{2\eta^2\beta}{c^2} + 1$. Since v should be a nonnegative and bounded function as $cy \to \pm\infty$, the only suitable solution is:

$$v(y) = e^{\frac{c}{2\eta}y}\left(C_- e^{-\frac{c\sqrt{D}}{4}y} + C_+ e^{\frac{c\sqrt{D}}{4}y}\right)^{-\frac{2}{\eta}} \tag{29}$$

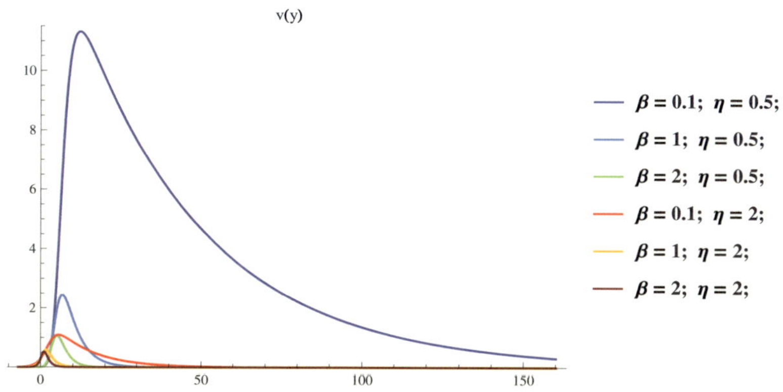

Figure 12.
$v(y)$; $c = 1$.

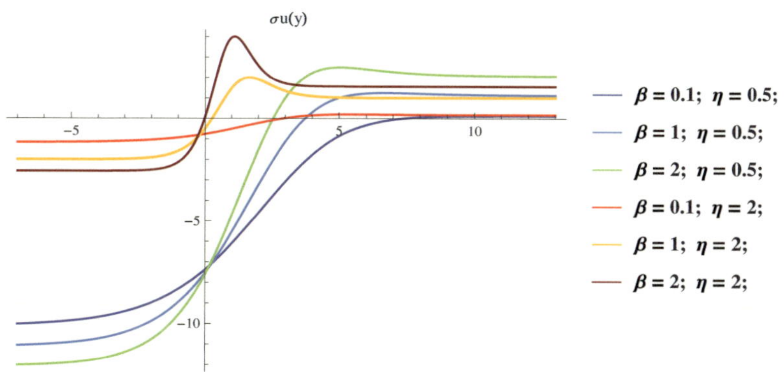

Figure 13.
$\sigma u(y)$; $c = 1$.

where $C_\pm$ are positive constants and $\beta > 0$. Unfortunately, the corresponding solution for $u(y)$ is alternating and has the form:

$$\begin{aligned} u(y) = &-\frac{\sigma c^2(\eta+2)}{2\eta^2}\left(C_-^2\left(1+\sqrt{D}\right)e^{-\frac{c\sqrt{D}}{4}y} + C_+^2\left(1-\sqrt{D}\right)e^{\frac{c\sqrt{D}}{4}y}\right. \\ &-\left.\frac{4\eta^2\beta}{c^2}C_-C_+\right)\left(C_-e^{-\frac{c\sqrt{D}}{4}y} + C_+e^{\frac{c\sqrt{D}}{4}y}\right)^{-\frac{2}{\eta}}. \end{aligned} \tag{30}$$

It is easy to see that $\sigma u(y) \to \frac{c^2(\eta+2)}{2\eta^2}\left(-1 \pm \sqrt{D}\right)$ as $cy \to \pm\infty$. These functions are presented in **Figures 12** and **13**.

4. Linear sensitivity

Let us consider the system with linear function $\phi(v) \sim v$. When $f(u,v) = u - v$, the system is called the minimal chemotaxis model following the nomenclature of [26]. This model is often considered with $f(u,v) = \tilde{\sigma}u - \tilde{\beta}v$ ($\tilde{\sigma}$ and $\tilde{\beta}$ are constants), and it is studied in many papers. As was proved in [27, 28], the solutions of this system are global and bounded in time for one space dimension. The case of positive $\tilde{\sigma}$ and nonnegative $\tilde{\beta}$ is studied in [29–33]. As we noted earlier, a sign of $\tilde{\sigma}$ may effect on the mathematical properties of the system, which changes its solvability conditions [34].

Now we consider the linear chemosensitivity function $\phi(v) = v$ and $f(u,v) = \tilde{\sigma}u - \tilde{\beta}v$. The replacement of $t \to \delta_1 t$, $v \to \frac{\eta_1}{\delta_1}v$, and $u \to \sigma\frac{\tilde{\sigma}\eta_1}{\delta_1^2}u$ leads to $\delta_1 = \eta_1 = 1$, $\alpha = \frac{\delta_2}{\delta_1}$, $\beta = \frac{\tilde{\beta}}{\delta_1}$, and $\sigma = \pm 1$. Then the system has the form:

$$\begin{cases} u_t - u_{xx} + (uv_x)_x = 0 \\ v_t - \alpha v_{xx} + \beta v - \sigma u = 0. \end{cases} \tag{31}$$

This system reduces to the system of ODEs in terms of traveling wave variable $y = x - ct$, where $c = const$:

$$\begin{cases} u_y + cu - uv_y + \lambda = 0 \\ \alpha v_{yy} + cv_y - \beta v + \sigma u = 0, \end{cases} \tag{32}$$

where $u = u(y)$, $v = v(y)$, and λ is an integration constant. As shown in [35], this system passes the Painlevé ODE test only if $\alpha = 2$ and $\beta = 0$. Let us focus on this case.

It is convenient to solve Eq. (32) in terms of variable:

$$z = \frac{\kappa}{|c|}e^{-\frac{cy}{2}}, \tag{33}$$

where $\kappa > 0$ is an arbitrary constant. Then for v and u, we obtain the solutions in the form:

$$\begin{aligned} v &= -\ln\left[\frac{|c|}{\kappa}zZ_\nu^2(z)\right] \\ u &= c^2z^2\left(1 - \frac{1}{4}(v_z)^2\right) - \frac{\lambda}{c}, \text{ where } \nu^2 = \frac{1}{4} - \frac{\lambda}{c^3}. \end{aligned} \tag{34}$$

The function $Z_\nu(z)$ satisfies the modified Bessel's equation and can be present as a linear combination of Infeld's and Macdonald's functions.

Using the series expansion of the Infeld's function, as well as theirs asymptotic behavior [36], one may obtain the following asymptotic forms for $e^{v_\nu(z)}$ and $u_\nu(z)$:

$$z \to \infty: \quad e^{v_\nu(z)} \to 0; \; u_\nu(z) \to 0. \tag{35}$$

$$z \to 0: \quad e^{v_\nu(z)} \to \begin{cases} \infty, & 0 \le \nu < \frac{1}{2}; \\ \frac{\kappa}{|c|C^2}\frac{8\pi}{(\pi+2)^2}, & \nu = \frac{1}{2}; \\ 0, & \nu > \frac{1}{2}; \end{cases} \tag{36}$$

$$u_\nu(z) \to c^2\left(\nu - \frac{1}{2}\right), \tag{37}$$

where the expression for $\nu = \frac{1}{2}$ agrees with Eq. (39).

So, the exact solution obtained has the form:

$$\begin{aligned} v &= -\ln\left[e^{-\frac{cy}{2}}A^2\left(I_\nu\left(\frac{\kappa}{|c|}e^{-\frac{cy}{2}}\right) + BK_\nu\left(\frac{\kappa}{|c|}e^{-\frac{cy}{2}}\right)\right)^2\right] \\ u &= -\sigma\left((v_y)^2 - \kappa^2 e^{-cy} + \frac{\lambda}{c}\right), \text{ where } \nu^2 = \frac{1}{4} - \frac{\lambda}{c^3}, \end{aligned} \tag{38}$$

where $\kappa > 0$, A, and B are arbitrary constants and the functions I_ν and K_ν are Infeld's and Macdonald's functions, respectively. This solution is not satisfactory from the biological point of view, since $v(y)$ is an alternating function for any ν. However it seems interesting because of the following: in the case of $\nu = \frac{1}{2}$ and $B = \frac{2+\pi}{2\pi}$ in terms of $e^{-\frac{cy}{2}}$, its form coincides with the well-known Korteweg-de Vries soliton.

Consider now the class of solutions with half-integer index $\nu = n + \frac{1}{2}$, when $Z_\nu(z)$ can be expressed in hyperbolic functions. The requirement of absence of divergence $u \to -\infty$ for finite z leads to the following form for $Z_{n+\frac{1}{2}}(z)$:

$$Z_{n+\frac{1}{2}}(z) = \begin{cases} Cz^{n+\frac{1}{2}}\left(\frac{d}{zdz}\right)^n \frac{\cosh(z+\zeta)}{z}, & n = 2k, \\ Cz^{n+\frac{1}{2}}\left(\frac{d}{zdz}\right)^n \frac{\sinh(z+\zeta)}{z}, & n = 2k+1; \; k = 0, 1 \dots; \\ \zeta = \frac{1}{2}\ln\frac{2}{\pi}, \; C = const. \end{cases} \tag{39}$$

At first let us consider the solutions obtained for $e^{v_{n+\frac{1}{2}}}$ and $u_{n+\frac{1}{2}}$ as functions of z. We begin with $n = 0$ or $\nu = \frac{1}{2}$. It is interesting to present the expressions for $e^{u_{\frac{1}{2}}(z)}$ and $u_{\frac{1}{2}}(z)$:

$$e^{v_{\frac{1}{2}}(z)} = \frac{\kappa}{C^2|c|}sech^2(z+\zeta) \tag{40}$$

$$u_{\frac{1}{2}}(z) = z^2c^2 sech^2(z+\zeta), \tag{41}$$

where Eq. (40) appears the one-soliton solution exactly the same as the well-known one of the Korteweg-de Vries equation. Returning to the variable y:

$$
\begin{aligned}
e^{v\left(e^{-\frac{cy}{2}}\right)} &= \frac{\kappa}{C^2|c|} sech^2\left(\frac{\kappa}{|c|} e^{-\frac{cy}{2}} + \frac{1}{2}\ln\frac{2}{\pi}\right) \\
u(y) &= \frac{\sigma(\pi B - 1)\kappa^2 e^{-cy}}{\left(\sinh\left(\frac{\kappa}{|c|} e^{-\frac{cy}{2}}\right) + \frac{\pi}{2} B e^{-\frac{\kappa}{|c|} e^{-\frac{cy}{2}}}\right)^2}.
\end{aligned} \tag{42}
$$

One can see that for $\sigma = 1$ (an increase of a chemical substance), the cell density $u(y) \geq 0$ for $B \geq \frac{1}{\pi}$ and that for $B > 0$ $u(y)$ is the solitary continuous solution vanishing as $y \to \pm\infty$, whereas for $B < 0$ $u(y)$ has a point of discontinuity. One can say that when $B < 0$, we obtain "blow-up" solution in the sense that it goes to infinity for finite y, and this is true for different ν.

The expressions for $n \geq 1$ become more complicated, and one can see the solitonic behavior of $e^{v_{n+\frac{1}{2}}(z)}$ and the curves for $u_{n+\frac{1}{2}}(z)$ in **Figures 14** and **15**.

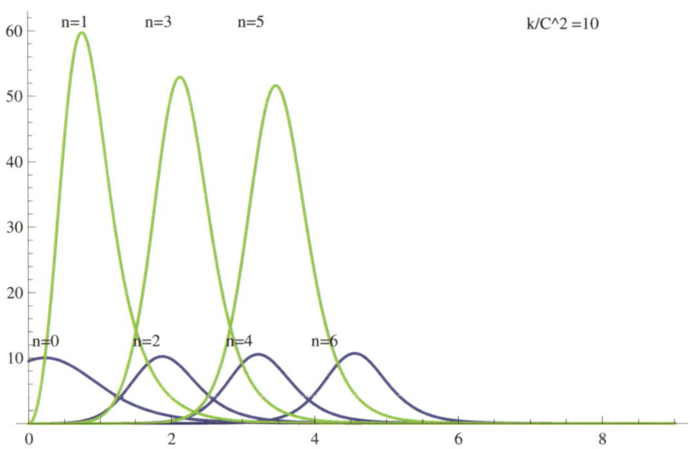

Figure 14.
$e^{v_{n+\frac{1}{2}}(z)}$; $n = 0, \ldots 6$; $c = 1$.

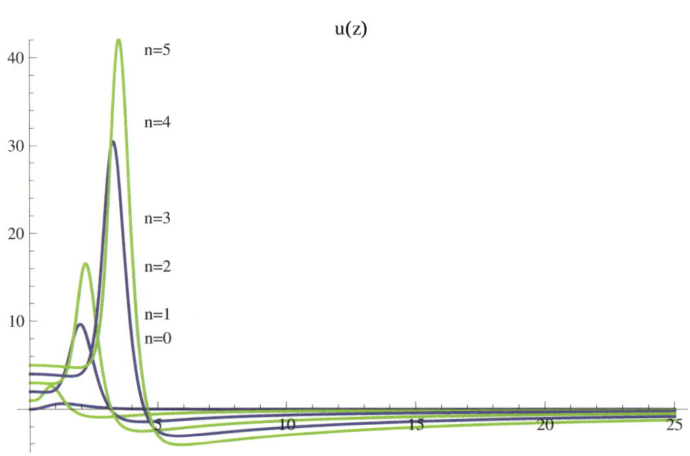

Figure 15.
$u_{n+\frac{1}{2}}(z)$; $n = 0, \ldots 5$; $c = 1$.

The explicit form of our solution in terms of the variable y can be obtained by direct substitution of Eq. (33) into Eq. (39), where $\frac{\lambda}{c} = -c^2 n(n+1)$. The resulting formulae are complicated and slightly difficult for analytic analysis; it seems to be more convenient to present the plots.

For $n = 0$ in the function $e^{v_{\frac{1}{2}}(y)}$, we have the "step" whose altitude depends on the values of velocity c and arbitrary constant κ. One may see that these curves become higher and shift to the right with different rates for the rising κ. The $u_{\frac{1}{2}}(y)$ is the positive function whose altitude and sharpness of peak depend on c (see **Figures 16** and **17**).

For $n \geq 1$ we can see that the solitonic behavior of $e^{v_{n+\frac{1}{2}}(y)}$ is retained for different values of c and κ; the curves become higher and more tight, and they shift to the right also with an increase of c and κ. For the cell density $u_{n+\frac{1}{2}}(y)$, the obtained solution has the negative section converging to zero for $cy \to -\infty$ (**Figures 18–21**).

The curves for the concentration of the chemical substance $v_{n+\frac{1}{2}}(y)$ are presented in **Figure 22**. Since $v_{n+\frac{1}{2}}(y)$ has to be positive (nonnegative), we see that these functions do not satisfy this requirement in all domains of definition.

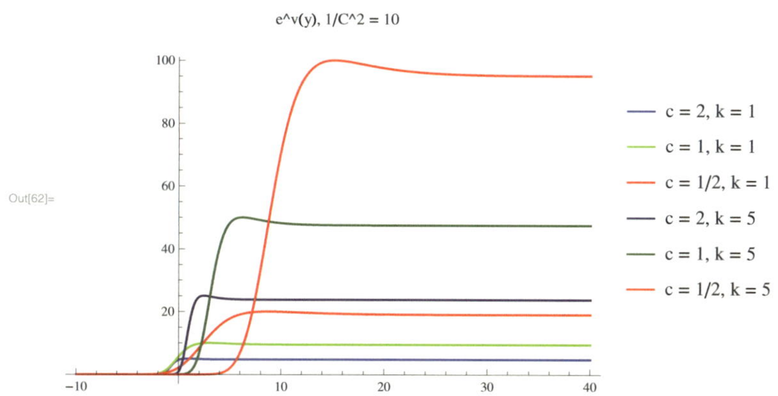

Figure 16.
$e^{v_{n+\frac{1}{2}}(y)}$; $n = 0$.

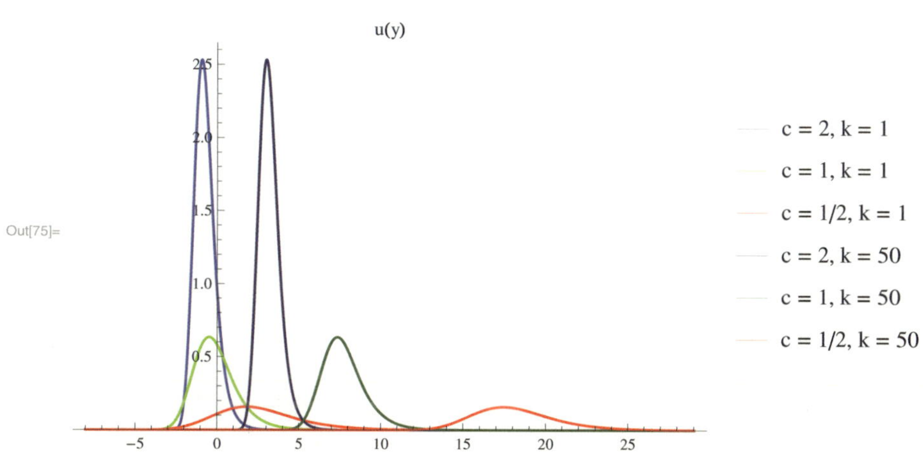

Figure 17.
$u_{n+\frac{1}{2}}(y)$; $n = 0$.

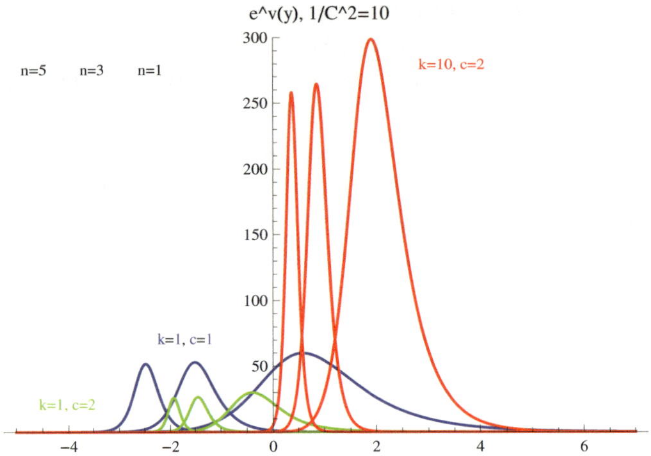

Figure 18.
$e^{v_{n+\frac{1}{2}}(y)}$; $n = 1;3;5$.

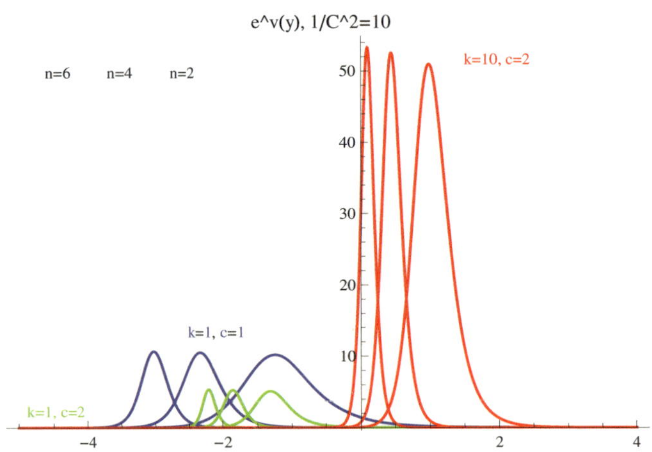

Figure 19.
$e^{v_{n+\frac{1}{2}}(y)}$; $n = 2;4;6$.

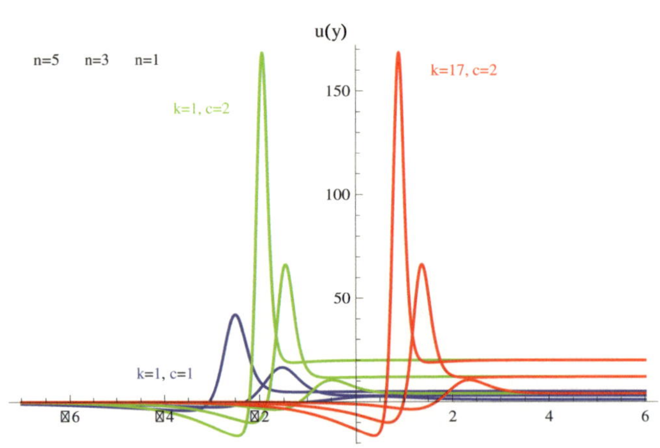

Figure 20.
$u_{n+\frac{1}{2}}(y)$; $n = 1;3;5$.

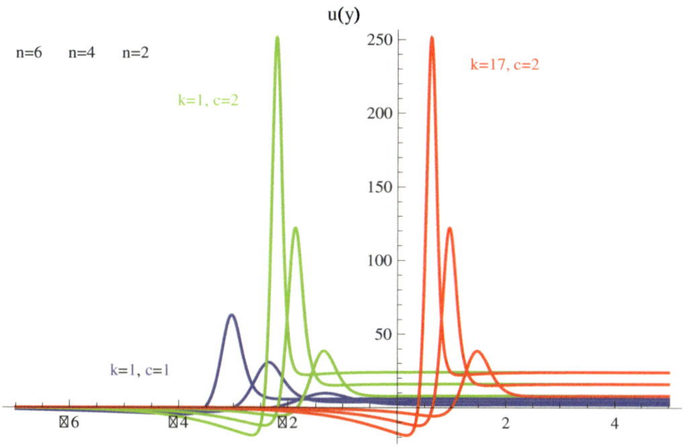

Figure 21.
$u_{n+\frac{1}{2}}(y); n = 2; 4; 6.$

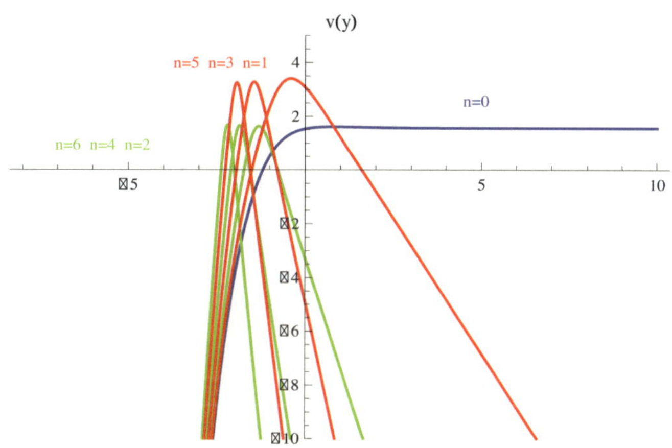

Figure 22.
$v_{n+\frac{1}{2}}(y); n = 0, 1, \ldots 6.$

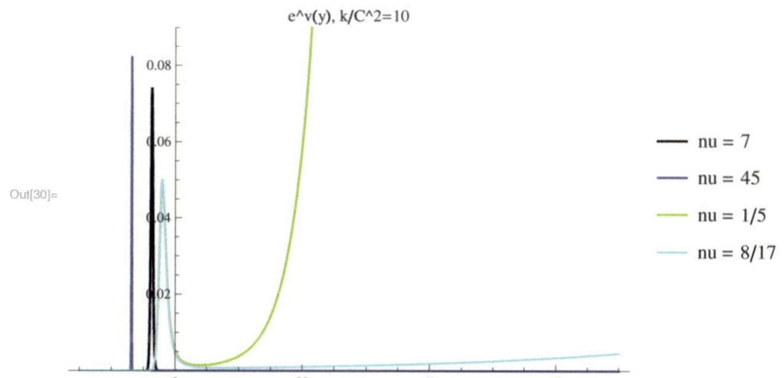

Figure 23.
$e^{v_\nu(y)}; \nu = 1/5; 8/17; 7; 45.$

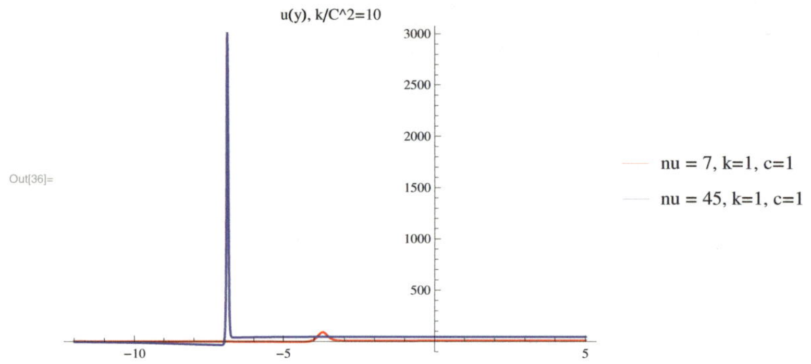

Figure 24.
$u_\nu(y)$; $\nu = 7; 45$.

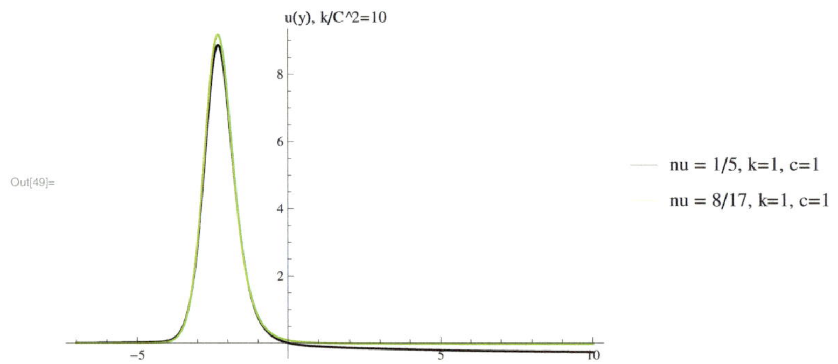

Figure 25.
$u_\nu(y)$; $\nu = 1/5; 8/17$.

In conclusion it seems interesting to present the plots for $e^{v_\nu(y)}$ and $u_\nu(y)$ for different values of ν (**Figures 23–25**). It is interesting to see that there are irregular solutions for $e^{v_\nu(y)}$; however, the corresponding solutions for $u_\nu(y)$ are regular [see Eqs. (35)–(37)].

5. Conclusion

We investigate three different one-dimensional parabolic-parabolic Patlak-Keller-Segel models. For each of them, we obtain the exact solutions in terms of traveling wave variables. Not all of these solutions are acceptable for biological interpretation, but there are solutions that require detailed analysis. It seems interesting to consider the latter for the experimental values of the parameters and see their correspondence with experiment. This question requires further investigations.

Author details

Maria Vladimirovna Shubina
Skobeltsyn Institute of Nuclear Physics, Lomonosov Moscow State University, Moscow, Russian Federation

*Address all correspondence to: yurova-m@rambler.ru

References

[1] Patlak CS. Random walk with persistence and external bias. The Bulletin of Mathematical Biophysics. 1953;**15**(3):311-338

[2] Keller EF, Segel LA. Initiation of slime Mold aggregation viewed as an instability. Journal of Theoretical Biology. 1970;**26**(3):399-415

[3] Keller EF, Segel LA. Model for Chemotaxis. Journal of Theoretical Biology. 1971;**30**(2):225-234

[4] Keller EF, Segel LA. Traveling bands of chemotactic bacteria: A theoretical analysis. Journal of Theoretical Biology. 1971;**30**(2):235-248

[5] Horstmann D. From 1970 until present: The Keller-Segel model in chemotaxis and its consequences. Preprints: Max-Planck-Institut für Mathematik in den Naturwissenschaften. Niedersächsische Staats- und Universitätsbibliothek. 2003

[6] Suzuki T. Free Energy and Self-Interacting Particles. Boston: Birkhuser; 2005

[7] Perthame B. Transport Equations in Biology. Basel: Birkhuser; 2007

[8] Hillen T, Painter KJ. A users guide to PDE models for chemotaxis. Journal of Mathematical Biology. 2009;**58**: 183-217

[9] Perthame B. PDE models for chemotactic movements: Parabolic, hyperbolic and kinetic. Applications of Mathematics. 2004;**49**(6):539-564

[10] Blanchet A, Dolbeault J, Perthame B. Two-dimensional Keller-Segel model: Optimal critical mass and qualitative properties of the solutions. Electronic Journal of Differential Equations. 2006;**44**:1-32

[11] Blanchet A, Laurenot P. The parabolic-parabolic Keller-Segel system with critical diffusion as a gradient flow in $\mathfrak{R}^d$, $d \geq 3$. Communications in Partial Differential Equations. 2013;**38**(4): 658-686

[12] Blanchet A, Carrillo JA, Kinderlehrer D, Kowalczyk M, Laurenot P, Lisini S. A hybrid variational principle for the Keller-Segel system in $\mathfrak{R}^2$. ESAIM: M2AN. 2015; **49**(6):1553-1576

[13] Ni W-M. Diffusion, cross-diffusion, and their spike-layer steady states. Notices of the American Mathematical Society. 1998;**45**(1):9-18

[14] Li T, Wang ZA. Nonlinear stability of large amplitude viscous shock waves of a generalized hyperbolic-parabolic system arising in chemotaxis. Mathematical Models and Methods in Applied Sciences. 2010;**20**:1967-1998

[15] Wang ZA. Mathematics of traveling waves in chemotaxis. Discrete and Continuous Dynamical Systems - Series B. 2013;**18**(3):601-641

[16] Nagai T, Ikeda T. Traveling waves in a chemotactic model. Journal of Mathematical Biology. 1991;**30**(2): 169-184

[17] Ebihara Y, Furusho Y, Nagai T. Singular solution of traveling waves in a chemotactic model. Bulletin of the Kyushu Institute of Technology, Mathematics, Natural Science. 1992;**39**: 29-38

[18] Bateman H, Erdélyi A. Higher Transcendental Functions. Vol. 2. New York, Toronto, London: McGHAW-HILL Book Company, INC; 1953

[19] Watson GN. A Treatise on the Theory of Bessel Functions. Cambridge, England: Cambridge University Press; 1944

[20] Olver PJ. Applications of Lie Groups to Differential Equations (Springer, 1986) (In translation). Moscow: Springer; 1989

[21] Zaitsev VF, Polyanin AD. Handbook on Ordinary Differential Equations. Moscow: Physmatlit; 2001

[22] Nossal R. Boundary movement of chemotactic bacterial populations. Mathematical Biosciences. 1972;**13**: 397-406

[23] Rosen G. Theoretical significance of the condition delta = 2mu in bacterial chemotaxis. Bulletin of Mathematical Biology. 1983;**45**(2):151-153

[24] Winkler M. Global solutions in a fully parabolic chemotaxis system with singular sensitivity. Mathematical Methods in the Applied Sciences. 2011; **34**:176-190

[25] Shubina M. Painlevé Analysis for Two 1D Parabolic-Parabolic Models of Chemotaxis: Some Travelling Wave Solutions. arXiv:1607.00349 [nlin.SI]

[26] Childress S, Percus JK. Nonlinear aspects of chemotaxis. Mathematical Biosciences. 1981;**56**:217-237

[27] Osaki K, Yagi A. Finite dimensional attractor for one-dimensional Keller-Segel equations. Funkcialaj Ekvacioj. 2001;**44**:441-469

[28] Hillen T, Potapov A. The one-dimensional chemotaxis model: Global existence and asymptotic profile. Mathematical Methods in the Applied Sciences. 2004;**27**:1783-1801

[29] Jäger W, Luckhaus S. On explosions of solutions to a system of partial differential equations modelling chemotaxis. Transactions of the American Mathematical Society. 1992; **329**:819-824

[30] Nagai T, Senba T, Yoshida K. Application of the Trudinger-Moser inequality to a parabolic system of chemotaxis. Funkcialaj Ekvacioj. 1997; **40**:411-433

[31] Corrias L, Escobedo M, Matos J. Existence, uniqueness and asymptotic behavior of the solutions to the fully parabolic Keller-Segel system in the plane. Journal of Differential Equations. 2014;**257**:1840-1878

[32] Tao Y, Winkler M. Boundedness in a quasilinear parabolic-parabolic Keller-Segel system with subcritical sensitivity. Journal of Differential Equations. 2012; **252**(1):692-715

[33] Fatkullin I. A study of blow-ups in the Keller-Segel model of chemotaxis. Nonlinearity. 2013;**26**(1):81-94

[34] Tupchiev VA, Fomina NA. On a correctness of two-dimensional boundary value problem for a system of equations of chemotaxis. Matematicheskoe Modelirovanie. 2001; **13**(12):95-106

[35] Shubina M. The 1D parabolic-parabolic Patlak-Keller-Segel model of chemotaxis: The particular integrable case and soliton solution. Journal of Mathematical Physics. 2016;**57**:091501

[36] Nikiforov AF, Uvarov VB. The Special Functions of Mathematical Physics. Moscow: Nauka; 1978

Chapter 2

Generalized and Fundamental Solutions of Motion Equations of Two-Component Biot's Medium

Lyudmila Alexeyeva and Yergali Kurmanov

Abstract

Here processes of wave propagation in a two-component Biot's medium are considered, which are generated by arbitrary forces actions. By using Fourier transformation of generalized functions, a fundamental solution, Green tensor, of motion equations of this medium has been constructed in a non-stationary case and in the case of stationary harmonic oscillation. These tensors describe the processes of wave propagation (in spaces of dimensions 1, 2, 3) under an action of power sources concentrated at coordinates origin, which are described by a singular delta-function. Based on them, generalized solutions of these equations are constructed under the action of various sources of periodic and non-stationary perturbations, which are described by both regular and singular generalized functions. For regular acting forces, integral representations of solutions are given that can be used to calculate the stress-strain state of a porous water-saturated medium.

Keywords: Biot's medium, solid and liquid components, Green tensor, Fourier transformation, regularization

1. Introduction

Various mathematical models of deformable solid mechanics are used to study the seismic processes of earth's crust. The processes of wave propagation are most studied in elastic media. But these models do not take into account many real properties of an ambient array. These are, for example, the presence of groundwater, which affects the magnitude and distribution of stresses. Models, which take into account the water saturation of earth's crust structures, presence of gas bubbles, etc., are multicomponent media. A variety of multicomponent media, complexity of processes associated with their deformation, lead to a large difference in methods of analysis and modeling used in solving such problems.

Porous medium saturated with liquid or gas, from the point of view of continuum mechanics, is essentially a two-phase continuous medium, one phase of which is particles of liquid (gas) and other solid particles are its elastic skeleton. There are various mathematical models of such media, developed by various authors. The most famous of them are the models of Biot, Nikolaevsky, and Horoshun [1–5]. However, the class of solved tasks to them is very limited and mainly associated with the construction and study of particular solutions of these equations based on methods of full and partial separation of variables and theory of special functions in the works of Rakhmatullin, Saatov, Filippov, Artykov [6, 7], Erzhanov, Ataliev, Alexeyeva, Shershnev [8, 9], etc. In this regard, it is important to develop effective methods of solution of boundary value problems for such media with the use of modern mathematical methods.

Periodic on time processes are very widespread in practice. By this cause, here we consider also processes of wave propagation in Biot's medium, posed by the periodic forces of different types. Based on the Fourier transformation of generalized functions, we constructed fundamental solutions of oscillation equations of Biot's medium. It is Green tensor, which describes the process of propagation of harmonic waves at a fixed frequency in the space–time of dimension N = 1, 2, 3, under the action concentrated at the coordinates origin. By using this tensor, we construct generalized solutions of these equations for arbitrary sources of periodic disturbances, which can be described as both regular and singular distributions. They can be used to calculate the stress-strain state of a porous water-saturated medium by seismic wave propagation.

2. The parameters and motion equations of a two-component Biot's medium

The equations of motion of a homogeneous isotropic two-component Biot's medium are described by the following system of second-order hyperbolic equations [1–3]:

$$\begin{gathered}(\lambda+\mu)\operatorname{grad}\operatorname{div}u_s+\mu\Delta u_s+Q\operatorname{grad}\operatorname{div}u_f+F^s(x,t)=\rho_{11}\ddot{u}_s+\rho_{12}\ddot{u}_f\\ Q\operatorname{grad}\operatorname{div}u_s+R\operatorname{grad}\operatorname{div}u_f+F^f(x,t)=\rho_{12}\ddot{u}_s+\rho_{22}\ddot{u}_f\\ (x,t)\in R^N\times[0,\infty).\end{gathered}\tag{1}$$

Here N is the dimension of the space. At a plane deformation $N = 2$, the total spatial deformation corresponds to $N = 3$, at $N = 1$ the equations describe the dynamics of a porous liquid-saturated rod.

We denote $u_s = u_{sj}(x,t)e_j$ is a displacement vector of an elastic skeleton, $u_f = u_{fj}(x,t)e_j$ is a displacement vector of a liquid, and $e_j(\,j = 1, \ldots, N)$ are basic orts of Lagrangian Cartesian coordinate system (everywhere by repeating indices, there is summation from 1 to N).

Constants $\rho_{11}, \rho_{12}, \rho_{22}$ have the dimension of mass density, and they are associated with densities of masses of particles, composing a skeleton $\rho_s$and a fluid ρ_f, by relationships:

$$\rho_{11}=(1-m)\rho_s-\rho_{12},\quad \rho_{22}=m\rho_f-\rho_{12},$$

where m is a porosity of the medium. The constant of attached density ρ_{12} is related to a dispersion of deviation of micro-velocities of fluid particles in pores from average velocity of fluid flow and depends on pores geometry. Elastic constants λ, μ are Lama's parameters of an isotropic elastic skeleton, and Q, R characterize an interaction of a skeleton with a liquid on the basis of.

2.1 Biot's law for stresses

$$\begin{gathered}\sigma_{ij}=\left(\lambda\partial_k u_{sk}+Q\partial_k u_{fk}\right)\delta_{ij}+\mu\left(\partial_i u_{sj}+\partial_j u_{si}\right)\\ \sigma=-mp=R\partial_k u_{fk}+Q\partial_k u_{sk}\end{gathered}\tag{2}$$

Here $\sigma_{ij}(x,t)$ are a stress tensor in a skeleton, and $p(x,t)$ is a pressure in a fluid component. External mass forces acting on a skeleton $F^s = F^s_j(x,t)e_j$ and on a liquid component $F^f = F^f_j(x,t)e_j$.

Further we use the next notations for partial derivatives: $\partial_k = \frac{\partial}{\partial x_k}$, $u_{j,k} = \partial_k u_j$, $\Delta = \partial_k\partial_k$ is Laplace operator.

There are three sound speeds in this medium:

$$c_1^2 = \frac{\alpha_1 + \sqrt{\alpha_1{}^2 - 4\alpha_2\alpha_3}}{2\alpha_2},$$
$$c_2^2 = \frac{\alpha_1 - \sqrt{\alpha_1{}^2 - 4\alpha_2\alpha_3}}{2\alpha_2}, \quad (3)$$
$$c_3^2 = \sqrt{\frac{\rho_{22}\mu}{\alpha_2}}$$

where the next constants were introduced as:

$$\alpha_1 = (\lambda + 2\mu)\rho_{22} + R\rho_{11} - 2Q\rho_{12},$$
$$\alpha_2 = \rho_{11}\rho_{22} - (\rho_{12})^2,$$
$$\alpha_3 = (\lambda + 2\mu)R - Q^2.$$

The first two speeds $c_1, c_2 (c_1 > c_2)$ describe the velocity of propagation of two types of *dilatational waves*. The second slower dilatation wave is called *repackaging wave*. A third velocity c_3 corresponds to *shear waves* and at $\rho_{12} = 0$ coincides with velocity of shear wave propagation in an elastic skeleton $(c_3 < c_1)$.

We introduce also two velocities of propagation of dilatational waves in corresponding elastic body and in an ideal compressible fluid:

$$c_s = \sqrt{\frac{\lambda + 2\mu}{\rho_{11}}}, \qquad c_f = \sqrt{\frac{R}{\rho_{22}}}$$

3. Problems of periodic oscillations of Biot's medium

Construction of motion equation solutions by periodic oscillations is very important for practice since existing power sources of disturbances are often periodic in time and therefore can be decomposed into a finite or infinite Fourier series in the form:

$$F^s(x,t) = \sum_n F^s_n(x)e^{-i\omega_n t},$$
$$F^f(x,t) = \sum_n F^f_n(x)e^{-i\omega_n t} \quad (4)$$

where periods of oscillation of each harmonic $T_n = 2\pi/\omega_n$ are multiple to the general period of oscillation T. Therefore, it is enough to consider the case of stationary oscillations, when the acting forces are periodic on time with an oscillation frequency ω:

$$F^s(x,t) = F^s(x)e^{-i\omega t},$$
$$F^f(x,t) = F^f(x)e^{-i\omega t} \quad (5)$$

The solution of Eq. (1) can be represented in the similar form:

$$u_s(x,t) = u_s(x)e^{-i\omega t}, \quad u_f(x) = u_f(x)e^{-i\omega t} \tag{6}$$

where complex amplitudes of displacements $u_s(x), u_f(x)$ must be determined. If the solution has been known for any frequency ω, then we get similar decomposition for displacements of a medium:

$$\begin{aligned} u_s(x,t) &= \sum_n u_{sn}(x)e^{-i\omega_n t}, \\ u_f(x,t) &= \sum_n u_{fn}(x)e^{-i\omega_n t} \end{aligned} \tag{7}$$

which give us the solution of problem for forces (4).

We get equations for complex amplitudes by stationary oscillations, substituting (6) into the system (1):

$$\begin{aligned} (\lambda+\mu)\mathrm{grad}\,\mathrm{div}\,u_s + \mu\Delta u_s + Q\,\mathrm{grad}\,\mathrm{div}\,u_f + \rho_{11}\omega^2 u_s + \rho_{12}\omega^2 u_f + F^s(x) = 0 \\ Q\,\mathrm{grad}\,\mathrm{div}\,u_s + R\,\mathrm{grad}\,\mathrm{div}\,u_f + \rho_{12}\omega^2 u_s + \rho_{22}\omega^2 u_f + F^f(x) = 0 \end{aligned} \tag{8}$$

To construct the solutions of this system for different forces, we define Green tensor of it.

4. Green tensor of Biot's equations by stationary oscillations

Let us construct $U_m^j(x,\omega)e^{-i\omega t}$ $(j,m = 1, \dots, 2N)$ fundamental solutions of the system (1) for the forces in the form:

$$F(x,t) = \begin{pmatrix} F^s \\ F^f \end{pmatrix} = \begin{pmatrix} \delta_k^{[j]} e_k \\ \delta_{k+N}^{[j]} e_k \end{pmatrix} \delta(x)e^{-i\omega t}, \tag{9}$$
$$k = 1, \dots, N, j = 1, \dots, 2N.$$

Here $\delta_k^j = \delta_{jk}$ is the Kronecker symbol, and $\delta(x)$ is the singular delta-function. They describe a motion of Biot's medium at an action of sources of stationary oscillations, concentrated in the point $x = 0$. The upper index of this tensor $(\dots^{[k]})$ fixes the current concentrated force and its direction. The lower index corresponds to component of movement of a skeleton and a fluid, respectively, $k = 1, \dots, N$ and $k = N+1, \dots, 2N$.

Their complex amplitudes $U_m^j(x,\omega)$ $(j,m = 1, \dots, 2N)$ satisfy the next system of equation:

$$\begin{aligned} &(\lambda+\mu)U^k_{j,ji} + \mu U^k_{i,jj} + \omega^2\rho_{11}U_i^k + QU^{k+N}_{j,ji} - \omega^2\rho_{12}U_i^{k+N} + \delta(x)\delta_j^k = 0 \\ &QU^k_{j,ji} + \rho_{12}\omega^2 U^k_{i\,,tt} + RU^{k+N}_{j,ji} + \rho_{22}\omega^2 U_i^{k+N} + \delta(x)\delta^k_{j+N} = 0 \\ &j = 1, \dots, 2N, \quad k = 1, \dots, 2N. \end{aligned} \tag{10}$$

Since fundamental solutions are not unique, we'll construct such, which tend to zero at infinity:

$$U_i^j(x,\omega) \to 0 \text{ at } \|x\| \to \infty \tag{11}$$

and satisfy the radiation condition of type of Sommerfeld radiation conditions [10].
Matrix of such fundamental equations is named *Green tensor* of Eq. (8).

5. Fourier transform of fundamental solutions

To construct $U_m^j(x,\omega)$, we use Fourier transformation, which for regular functions has the form:

$$F[\varphi(x)] = \overline{\varphi}(\xi) = \int_{R^N} \varphi(x) e^{i(\xi,x)} dx_1 \dots dx_N$$

$$F^{-1}[\overline{\varphi}(\xi)] = \varphi(x) = \frac{1}{(2\pi)^N} \int_{R^N} \overline{\varphi}(\xi) e^{-i(\xi,x)} d\xi_1 \dots d\xi_N$$

where $\xi = (\xi_1, \dots, \xi_N)$ are Fourier variables.

Let us apply Fourier transformation to Eq. (10) and use property of Fourier transform of derivatives [10]:

$$\frac{\partial}{\partial x_j} \leftrightarrow -i\xi_j \tag{12}$$

Then we get the system of 2 N linear algebraic equations for Fourier components of this tensor:

$$\begin{aligned} &-(\lambda+\mu)\xi_j\xi_j\overline{U}_j^k - \mu\|\xi\|^2\overline{U}_j^k - Q\xi_j\xi_j\overline{U}_{j+N}^k + \rho_{11}\omega^2\overline{U}_j^k + \rho_{12}\omega^2\overline{U}_{j+N}^k + \delta_j^k = 0 \\ &-Q\xi_j\xi_j\overline{U}_j^k - R\xi_j\xi_l\overline{U}_{j+N}^k + \rho_{12}\omega^2\overline{U}_j^k + \rho_{22}\omega^2\overline{U}_{j+N}^k + \delta_{j+N}^k = 0 \\ &j = 1, \dots, N, \qquad k = N+1, \dots, 2N \end{aligned} \tag{13}$$

By using gradient divergence method, this system has been solved by us. For this the next basic function were introduced:

$$\begin{aligned} f_{0k}(\xi,\omega) &= \frac{1}{c_k^2\|\xi\|^2 - \omega^2}, \\ f_{jk}(\xi,\omega) &= \frac{f_{(j-1)k}(\xi,\omega)}{-i\omega}, \quad j = 1, 2; \end{aligned} \tag{14}$$

and the next theorem has been proved [11, 12].

Theorem 1. *Components of Fourier transform of fundamental solutions have the form:*

$$for\ j = \overline{1,N}, \qquad k = \overline{1,N},$$

$$\overline{U}_j^k = \left(-i\xi_j\right)(-i\xi_k)[\beta_1 f_{21} + \beta_2 f_{22} + \beta_3 f_{23}] + $$

$$+\frac{1}{\alpha_2}\left(\rho_{12}\delta_{j+N}^k - \rho_{22}\delta_j^k\right)f_{03},$$

$$\overline{U}_{j+N}^k = \left(-i\xi_j\right)(-i\xi_k)[\gamma_1 f_{21} + \gamma_2 f_{22} + \gamma_3 f_{23}] -$$

$$-\frac{\mu}{\alpha_2}\delta_{j+N}^k\|\xi\|^2 f_{23} - \frac{1}{\alpha_2}\left(\rho_{11}\delta_{j+N}^k + \rho_{12}\delta_j^k\right)f_{03};$$

$$for \quad j=1,\ldots,N \qquad k=N+1,\ldots,2N$$

$$\overline{U}_j^k = \left(-i\xi_j\right)(-i\xi_{k-N})\left[\eta_1 f_{21}+\eta_2 f_{22}+\eta_3 f_{23}\right]+$$

$$+\frac{1}{\alpha_2}\left(\rho_{12}\delta^k_{j+N}-\rho_{22}\delta^k_j\right)f_{03}$$

$$\overline{U}^k_{j+N} = \left(-i\xi_j\right)(-i\xi_{k-N})[\varsigma_1 f_{21}+\varsigma_2 f_{22}+\varsigma_3 f_{23}]-$$

$$-\frac{\mu}{\alpha_2}\delta^k_{j+N}\|\xi\|^2 f_{23}-\frac{1}{\alpha_2}\left(\rho_{11}\delta^k_{j+N}+\rho_{12}\delta^k_j\right)f_{03}$$

where the next constants have been introduced as:

$$D_1=\frac{1}{\alpha_2 \upsilon_{12}}, \quad \upsilon_{lm}=c_l^2-c_m^2, q_1=Q\rho_{12}-(\lambda+\mu)\rho_{12}, \quad q_2=\rho_{11}R-Q\rho_{12},$$

$$d_1=(\lambda+\mu)\rho_{22}-Q\rho_{12}, \quad d_2=Q\rho_{22}-R\rho_{12}, \quad d_{3j}=\rho_{12}c_j^2-Q \quad (j=1,2)$$

$$\beta_j=(-1)^{(j+1)}\frac{D_1c_j^2}{\alpha_2\upsilon_{3j}}(d_1b_{sj}+d_2d_{3j}), \quad \beta_3=-\frac{c_3^2}{\alpha_2\upsilon_{31}\upsilon_{32}}(d_1b_{s3}+d_2d_{33});$$

$$\gamma_j=(-1)^{j+1}\frac{D_1c_j^2}{\alpha_2\upsilon_{3j}}(q_1b_{fj}+q_2d_{3j}), \quad \gamma_3=-\frac{D_1c_3^2\upsilon_{12}}{\alpha_2\upsilon_{31}\upsilon_{32}}(q_1b_{f3}+q_2d_{33});$$

$$\eta_j=(-1)^{j+1}\frac{D_1c_j^2}{\alpha_2\upsilon_{3j}}(d_jd_{3j}+d_2b_{js}), \quad \eta_3=-\frac{c_3^2\upsilon_{12}}{\alpha_2\upsilon_{31}\upsilon_{32}}(d_1d_{33}+d_2b_{3s});$$

$$\varsigma_j=(-1)^{j+1}\frac{D_1c_j^2}{\alpha_2\upsilon_{3j}}(q_1d_{3j}+q_2b_{(4-j)s}), \quad \varsigma_3=-\frac{c_3^2\upsilon_{12}}{\alpha_2\upsilon_{31}\upsilon_{32}}(q_1d_{33}+q_2b_{3s})$$

$$b_{fj}=\rho_{22}\upsilon_{fj}, b_{sj}=\rho_{11}\upsilon_{js}.$$

This form is very convenient for constructing originals of Green tensor.

6. Stationary Green tensor construction: radiation conditions

In this case let us construct the originals of the basic function but only over ξ by constant frequency:

$$\Phi_{0m}(x,\omega)=F_\xi^{-1}\left[f_{0m}(\xi,\omega)\right]$$

which, in accordance with its definition (14), satisfies the equation:

$$\left(c_m^2\|\xi\|^2-\omega^2\right)f_{0m}=1 \tag{15}$$

Using property (12) for derivatives from here, we get Helmholtz equation for fundamental solution (accurate within a factor c_m^{-2}):

$$(\Delta+k_m^2)\Phi_{0m}+c_m^{-2}\delta(x)=0, \quad k_m=\frac{\omega}{c_m} \tag{16}$$

Fundamental solutions of Helmholtz equation, which satisfy Sommerfeld conditions of radiation:

$$\text{at } r \to \infty$$

$$\Phi'_{0m}(r) - ik_m\Phi_{0m}(r) = O\left(r^{-1}\right), \qquad N = 3,$$

$$\Phi'_{0m}(r) - ik_m\Phi_{0m}(r) = O\left(r^{-1/2}\right), \quad N = 2.$$

are well known [10]. They are unique. Using them, we obtain:

for $N = 3$

$$\Phi_{0m} = \frac{1}{4\pi rc^2}e^{ik_m r}, \quad k_m = \frac{\omega}{c_m};$$

for $N = 2$

$$\Phi_{0m} = \frac{i}{4c^2}H_0^{(1)}(k_m r),$$

where $H_j^{(1)}(k_m r)$ is the cylindrical Hankel function of the first kind:

for $N = 1$

$$\Phi_{0m} = \frac{\sin k_m|x|}{2k_m c_m^2}.$$

These functions (subject to factor $e^{-i\omega t}$) describe harmonic waves which move from the point $x = 0$ to infinity and decay at infinity.

The last property is true only for $N = 2{,}3$. In the case $N = 1$, all fundamental solutions of Eq. (16):

$$\left(\frac{d^2}{dx^2} + k_m^2\right)\Phi_{0m} + c_m^{-2}\delta(x) = 0,$$

do not decay at infinity.

From Theorem 1, the next theorem follows.

Theorem 2. *The components of Green tensor of Biot's equations at stationary oscillations with frequency ω, which satisfy the radiation conditions, have the form:*

for $j = \overline{1,N}, \quad k = \overline{1,N},$

$$U_j^k(x,\omega) = -\omega^{-2}\sum_{m=1}^{3}\beta_m\frac{\partial^2\Phi_{0m}}{\partial x_j\partial x_k} + \frac{1}{\alpha_2}\left(\rho_{12}\delta^k_{j+N} - \rho_{22}\delta_j^k\right)\Phi_{03},$$

$$U^k_{j+N}(x,\omega) = -\omega^{-2}\sum_{m=1}^{3}\gamma_m\frac{\partial^2\Phi_{0m}}{\partial x_j\partial x_k} +$$

$$+\frac{\mu\delta^k_{j+N}}{\alpha_2\omega^2}\left(c_3^{-2}\delta(x) + k_3^2\Phi_{0m}\right) - \frac{\rho_{11}\delta^k_{j+N} + \rho_{12}\delta_j^k}{\alpha_2}\Phi_{03};$$

for $j = 1, \ldots, N \quad k = N+1, \ldots, 2N$

$$U_j^k(x,\omega) = -\omega^{-2}\sum_{m=1}^{3}\eta_m\frac{\partial^2\Phi_{0m}}{\partial x_j\partial x_k} + \frac{1}{\alpha_2}\left(\rho_{12}\delta^k_{j+N} - \rho_{22}\delta^k_j\right)\Phi_{03}$$

$$U^k_{j+N}(x,\omega) = -\omega^{-2}\sum_{m=1}^{3}\varsigma_m\frac{\partial^2\Phi_{0m}}{\partial x_j\partial x_k} +$$

$$+\frac{\mu}{\alpha_2\omega^2}\left(c_3^{-2}\delta(x) + k_3^2\Phi_{0m}\right)\delta^k_{j+N} - \frac{1}{\alpha_2}\left(\rho_{11}\delta^k_{j+N} + \rho_{12}\delta^k_j\right)\Phi_{03}$$

where

for $N = 1$

$$\frac{d^2\Phi_{0m}}{dx^2} = \frac{1}{2c_m^2k_m}\left({k_m}^2(\sin k_m|x|) - 2k_m\delta(x)\right);$$

for $N = 2$

$$\frac{\partial^2\Phi_{0m}}{\partial x_j\partial x_k} = -\frac{i}{4c_m^2}\left(0.5k_m^2(H_0(k_mr) - H_2(k_mr))r,_jr,_k + kH_1^1(k_mr)r,_{jk}\right);$$

for $N = 3$

$$\frac{\partial\Phi_{0m}}{\partial x_j\partial x_k} = \frac{1}{4\pi rc_m^2}e^{ikr}\left\{r,_jr,_k\left(\left(ik_m - \frac{1}{r}\right)^2 + \frac{1}{r^2}\right) + r,_{jk}\left(ik_m - \frac{1}{r}\right)\right\};$$

$$k_m = \frac{\omega}{c_m}, r = \|x\|,\quad r,_j = \frac{x_j}{r},\quad r,_{ij} = \frac{1}{r}\left(\delta_{ij} - \frac{x_ix_j}{r^2}\right)..$$

Proof. By using originals of basic functions, property (12) of derivatives, we can obtain from formulas for U_j^k in Theorem 1 the originals of all addends, besides that which contain factor $\|\xi\|^2$. But using (16) we have:

$$-\Delta\Phi = c_m^{-2}\delta(x) + k_m^2\Phi_{0m} \quad \leftrightarrow \quad \|\xi\|^2f_{0m} = c_m^{-2} + k_m^2f_{0m}$$

Then formulas of Theorem 2 follow from formulas of Theorem 1.

7. Generalized solutions by arbitrary periodic forces

Under the action of arbitrary mass forces with frequency ω in Biot's medium, the solution for complex amplitudes has the form of a tensor functional convolution:

$$u_j(x,t) = U_j^k(x,\omega) * F_k(x)e^{-i\omega t}, j,k = \overline{1,2N} \tag{17}$$

Note that mass forces may be different from the space of generalized vector function, singular and regular. Since Green tensor is singular and contains delta-functions, this convolution is calculated on the rule of convolution in generalized function space. If a support of acting forces are bounded (contained in a ball of finite radius), then all convolutions exist. If supports are not bounded, then the existence conditions of convolutions in formula (17) requires some limitations on behavior of forces at infinity which depends on a type of mass forces and space dimension.

The obtained solutions allow us to study the dynamics of porous water- and gas-saturated media at the action of periodic sources of disturbances of a sufficiently arbitrary form. In particular, they are applicable in the case of actions of certain forces on surfaces, for example, cracks, in porous media that can be simulated by simple and double layers on the crack surface.

There is another interesting feature of the Green tensor of the Biot's equations, which contains, as one of the terms, the delta-function that complicates the application of this tensor for solving boundary value problems based on analogues of Green formulas for elliptic systems of equations or the boundary element method. Here, when constructing the model, the viscosity of the liquid is not taken into account, which, apparently, leads to the presence of such terms, and it requires improvement of this model taking into account a viscosity.

8. Green tensor of Biot's equations by non-stationary motion

To construct the non-stationary Green tensor, at first we also construct the originals of the basic functions in an initial space-time:

$$\mathbf{\Phi}_{0m}(x,t) = F^{-1}\left[f_{0m}(\xi,\omega)\right] = F^{-1}\left[\left(c_m^2\|\xi\|^2 - \omega^2\right)^{-1}\right]$$

They are originals of the classic wave equation:

$$\left(\frac{\partial^2}{\partial t^2} - c_m^2\Delta\right)\mathbf{\Phi}_{0m} = \delta(t)\delta(x) \tag{18}$$

Depending on the dimension of a space, solutions of this wave equation that satisfy the radiation conditions have the following form [10]:

$$\mathbf{\Phi}_{0m}(x,t) = \frac{1}{4\pi c_m^2 r}\delta\left(t - \frac{r}{c_m}\right), \quad N = 3; \tag{19}$$

$$\mathbf{\Phi}_{0m}(x,t) = \frac{1}{2\pi c_m}\frac{H(ct-r)}{\sqrt{c_m^2 t^2 - r^2}}, \quad N = 2; \tag{20}$$

$$\mathbf{\Phi}_{0m}(x,t) = \frac{1}{2}H(c_m t - |x|), \quad N = 1. \tag{21}$$

Here $H(t)$ is the Heaviside function, and singular function $\delta(t - r/c_m)$ is the *simple layer on the sound cone* $r = c_m t$, $r = \|x\|$.

Using regularization of the general function ω^{-1}in the space of distribution [10]:

$$H(t)\delta(x) \leftrightarrow \frac{1}{-i(\omega + i0)}$$

and the properties of Fourier transform of generalized functions convolution:

$$h = f * g \leftrightarrow \bar{h} = \bar{f} \times \bar{g}$$

It is easy to show that the next lemma is true.

Lemma. *The originals of the primitives of the basic functions satisfying the radiation conditions are representable in the following form:*

for N = 3

$$\mathbf{\Phi}_{1m}(x,t) = \mathbf{\Phi}_{0m}(x,t) * H(t)\delta(x) = \frac{H(c_m t - r)}{4\pi c^2 r},$$
$$\mathbf{\Phi}_{2m}(x,t) = \mathbf{\Phi}_{1m}(x,t) * H(t)\delta(x) = \frac{(c_m t - r)_+}{4\pi c^3 r}; \quad (22)$$

for N = 2

$$\mathbf{\Phi}_{1m}(x,t) = \frac{1}{2\pi c^2}\ln\left(\frac{c_m t + \sqrt{c_m^2 t^2 - r^2}}{r}\right),$$
$$\mathbf{\Phi}_{2m}(x,t) = \frac{1}{2\pi c^3}\left(c_m t \ln\left(\frac{c_m t + \sqrt{c_m^2 t^2 - r^2}}{r}\right) - \sqrt{c_m^2 t^2 - r^2}\right); \quad (23)$$

for N = 1

$$\mathbf{\Phi}_{1m}(x,t) = 0,5(c_m t - r)H(c_m t - r) \triangleq 0.5\,(c_m t - r)_+,$$
$$\mathbf{\Phi}_{2m}(x,t) = \frac{1}{2c^2}(c_m t - r)^2 H(c_m t - r) \triangleq \frac{1}{2c^2}(c_m t - r)_+^2. \quad (24)$$

Using these functions and the properties of the Fourier transform, we obtain the components of the Green tensor from the formulas of Theorem 1. We formulate the result in the next theorem.

Theorem 3. *The components of Green tensor of motion equations of two-component Biot's medium have the following forms:*

For $j = \overline{1,N}, \quad k = \overline{1,N},$

$$U_j^k(x,t) = \sum_{m=1}^{3} \beta_m \frac{\partial^2 \mathbf{\Phi}_{2m}}{\partial x_j \partial x_k} + \frac{1}{\alpha_2}\left(\rho_{12}\delta^k_{j+N} - \rho_{22}\delta_j^k\right)\mathbf{\Phi}_{03}(x),$$

$$U^k_{j+N}(x,t) = \sum_{m=1}^{3} \gamma_m \frac{\partial^2 \mathbf{\Phi}_{2m}}{\partial x_j \partial x_k} + \frac{\mu}{\alpha_2 c_m^2}\delta^k_{j+N}(\mathbf{\Phi}_{0m} - t_+\delta(x)) -$$
$$-\frac{1}{\alpha_2}\left(\rho_{11}\delta^k_{j+N} + \rho_{12}\delta_j^k\right)\mathbf{\Phi}_{03}(x)$$

For $j = \overline{1,N}, \quad k = \overline{N+1,2N},$

$$U_j^k(x,t) = \sum_{m=1}^{3} \eta_m \frac{\partial^2 \mathbf{\Phi}_{2m}}{\partial x_j \partial x_{k-N}} + \frac{1}{\alpha_2}\left(\rho_{12}\delta^k_{j+N} - \rho_{22}\delta_j^k\right)\mathbf{\Phi}_{03}(x),$$

$$U^k_{j+N}(x,t) = \sum_{m=1}^{3} \varsigma_m \frac{\partial^2 \mathbf{\Phi}_{2m}}{\partial x_j \partial x_{k-N}} + \frac{\mu}{\alpha_2 c_m^2}\delta^k_{j+N}(\mathbf{\Phi}_{0m} - t_+\delta(x)) -$$
$$-\frac{1}{\alpha_2}\left(\rho_{11}\delta^k_{j+N} + \rho_{12}\delta_j^k\right)\mathbf{\Phi}_{03}(x)$$

Here
for *N* = 1

$$\frac{d^2\mathbf{\Phi}_{2m}}{dx^2} = \frac{H(c_m t - |x|)}{c_m^2} - \frac{2}{c_m^2}(c_m t - |x|)_+\delta(x) \quad (25)$$

for $N = 2$

$$\frac{\partial^2 \Phi_{2m}}{\partial x_j \partial x_k} = \frac{H(c_m t - r)}{2\pi c_m^3 r^3} \left(\frac{2c_m^2 t^2 - r^2}{\sqrt{c_m^2 t^2 - r^2}} r,_k r,_j - \delta_{jk} \sqrt{c_m^2 t^2 - r^2} \right) \tag{26}$$

for $N = 3$

$$\frac{\partial^2 \Phi_{2m}}{\partial x_j \partial x_k} = \frac{t}{4\pi c_m^2 r^2} \left(\delta(c_m t - r) r,_k r,_j - \frac{tH(c_m t - r)}{r} \left(\delta_{jk} - 3r,_k r,_j \right) \right), \tag{27}$$

$$r,_j = x_j / r.$$

9. Generalized solutions of Biot's equations by non-stationary forces

Using the properties of Green tensor, we obtain generalized solutions of non-stationary Biot's equations under the action of arbitrary mass forces in the Biot's medium, which satisfy the radiation condition at infinity. They have the form of tensor functional convolution:

$$u_j(x,t) = U_j^k(x,t) * F_k(x,t), \quad j,k = 1, \ldots, 2N \tag{28}$$

It's taken according to the rules of convolution of generalized functions depending on the type of mass forces [10].

In order to get the classic solution, we must present formulas (28) in regular integral forms. For this, let us present matrix of Green tensor as sum of regular functions and singular functions, which contain delta-function:

$$U(x,t) = U_{reg}(x,t) + U_{sing}(t)\delta(x).$$

Then also write:

$$u(x,t) = u1(x,t) + u2(x,t) \tag{29}$$

Here $u1(x,t)$ is representable by regular mass forces in the integral form:

$$u1(x,t) = H(t) \int_0^t d\tau \int_{R^N} U_{reg}(x - y, \tau) \times \begin{pmatrix} F_s(y, t-\tau) \\ F_f(y, t-\tau) \end{pmatrix} dy$$

The convolution with singular part is equal to:

$$u2(x,t) = H(t) \int_0^t U_{sing}(\tau) \times \begin{pmatrix} F_s(x, t-\tau) \\ F_f(x, t-\tau) \end{pmatrix} d\tau$$

In 3D space, there are convolutions with simple layers on sound cones (see (27)). To construct their integral presentation, use this rule:

$$\alpha(x,t)\delta(c_m t - r) * F(x,t) =$$

$$= H(t) \int_0^t d\tau \int_{\|y-x\|=c_m\tau} \alpha(x-y,\tau) F\left(y, t - \frac{\|y-x\|}{c_m}\right) dS(y)$$

Here the internal integral is taken over sphere with center in the point x, and its radius is equal to $c_m\tau$.

If components of acting forces $F(x,t)$ are double differentiable vector function, it is convenient to use the property of differentiation of convolution [10]:

$$\frac{\partial^2 \mathbf{\Phi}_{2m}}{\partial x_j \partial x_k} * F(x,t) = \frac{\partial^2}{\partial x_j \partial x_k}(\mathbf{\Phi}_{2m} * F(x,t)) = \mathbf{\Phi}_{2m}(x,t) * \frac{\partial^2 F}{\partial x_j \partial x_k}$$

Substituting the formulas of Theorem 4 into (29), we obtain displacements and stresses of skeleton and liquid in Biot's medium in spaces of dimension N = 1, 2, 3. Calculation of these convolutions by using these formulas essentially depends on the form of acting forces and gives possibility to construct regular presentation of generalized solution for wide class of acting forces, which are the classic solution of Biot's equation.

10. Calculation of the stress state of Biot's medium

Using Biot's law (2), we can define the generalized stresses in skeleton and a pressure in a liquid:

$$\begin{aligned} \sigma_{ij} &= \left(\lambda \partial_l U_l^k * F_{ks} + Q \partial_l U_l^{k+N} * F_{kf}\right)\delta_{ij} + \\ &+ \mu\left(\partial_i U_j^k * F_{ks} + \partial_j U_i^{k+N} * F_{kf}\right) \\ \sigma &= -mp = R\partial_l U_l^{k+N} * F_{kf} + Q\partial_l U_l^k * F_{ks} \end{aligned} \tag{30}$$

These formulas also can be written in integral form by using the same rules. But we can apply here the next lemma, which was proved in [11].

Lemma. *Fourier transformations of the divergences of Green tensor have the next form:*

by $k = 1, \dots, N$

$$F\left[\partial_j U_j^k\right] = D_1 i\xi_k \left(b_{f1} f_{01}(\xi,\omega) - b_{f2} f_{02}(\xi,\omega)\right)$$

$$F\left[\partial_j U^k_{j+N}\right] = D_1 i\xi_k (d_{31} f_{01}(\xi,\omega) - d_{32} f_{02}(\xi,\omega))$$

$j = 1, \dots, N.$

$$by\ k = N+1, \dots, 2N$$

$$F\left[\partial_j U_j^k\right] = i\xi_{k-N} D_1\left(d_{31} f_{01}(\xi,\omega) - d_{32} f_{02}(\xi,\omega)\right) F\left[\partial_j U^k_{j+N}\right] = i\xi_{k-N} D_1 (b_{s1} f_{01}(\xi,\omega) - b_{s2} f_{02}(\xi,\omega))$$

From this lemma, we can prove easily the next theorem.

Theorem 4. *Divergences of elastic and liquid displacement of Green tensor have the next form:*

for $k = 1, \dots, N$

$$\partial_j U_j^k = -D_1\left(b_{f1} \partial_k \mathbf{\Phi}_{01} - b_{f2} \partial_k \mathbf{\Phi}_{02}\right)$$

$$\partial_j U^k_{j+N} = -D_1 (d_{31} \partial_k \mathbf{\Phi}_{01} - d_{32} \partial_k \mathbf{\Phi}_{02})$$

$j = 1, \dots, N.$

for $k = N + 1, \dots, 2N$

$$\partial_j U_j^k = -D_1 (d_{31}\partial_{k-N}\mathbf{\Phi}_{01} - d_{32}\partial_{k-N}\mathbf{\Phi}_{02})$$
$$\partial_j U^k_{j+N} = -D_1 (b_{s1}\partial_{k-N}\mathbf{\Phi}_{01} - b_{s2}\partial_{k-N}\mathbf{\Phi}_{02})$$
$$j = 1, \dots, N.$$

Substituting these formulas in (30), we define the stresses in the skeleton and the pressure in the liquid of Biot's medium.

If we paste $\Phi_{02}(x, \omega)$ instead of $\mathbf{\Phi}_{02}(x, t)$ in formulas of this theorem, then formula (30) expresses complex amplitudes of stress tensor and pressure by periodic oscillations. It is used to determine stresses and pressure by solving the periodic problems (4).

11. Conclusion

The obtained solutions give possibility to study the dynamics of porous water- and gas-saturated media and rods under actions of disturbance sources of different forms and can be used for solutions of boundary value problems in porous media by using boundary element method.

These solutions can be used for describing wave processes by explosions and earthquakes. In these cases mass forces are described by using singular generalized function, such as multipoles, simple and double layers, and others.

Acknowledgements

This work was financially supported by the Ministry of Education and Science of the Republic of Kazakhstan (Grant AP05132272).

Author details

Lyudmila Alexeyeva[1*] and Yergali Kurmanov[1,2]

1 Institute of Mathematics and Mathematical Modeling, Almaty, Kazakhstan

2 Al Farabi Kazakh National University, Almaty, Kazakhstan

*Address all correspondence to: alexeeva@math.kz

References

[1] Biot MA. Theory of propagation of elastic waves in a fluid-saturated porous solid. 1. Low frequency range. The Journal of the Acoustical Society of America. 1956;**28**(2):168-178

[2] Biot MA. Theory of propagation of elastic waves in a fluid-saturated porous solid. 2. Higher frequency range. The Journal of the Acoustical Society of America. 1956;**28**(2):178-191

[3] Biot MA. Mechanics of deformation and propagation of acoustic waves in a porous medium. Mechanics. 1963;**6**(28): 103-110

[4] Nikolayevsky VN. Mechanics of Porous and Fractured Media. Moscow: Nauka; 1984. p. 232

[5] Horoshun LP. To the theory of saturated porous media. International Journal of Applied Mechanics. 1976; **12**(12):36-48

[6] Rakhmatullin KA, Saatov YU, Filippov IG, Artykov TU. Waves in Two-Component Media. Tashkent: Nauka UzSSR; 1974. p. 266

[7] Saatov YU. Plane Problems of Mechanics of Elastic-Porous Media. Tashkent: Nauka UzSSR; 1972. p. 248

[8] Yerzhanov ZS, Aitaliev SM, Alexeyeva LA. Dynamics of Tunnels and Underground Pipelines. Alma-Ata: Nauka KazSSR; 1989. p. 240

[9] Alexeyeva LA, Shershnev VV. Fundamental solutions of the equations of motion equation of Biot are medium. In: Reports of National Academy of Sciences of Rep. Kazakhstan. Vol. 1. 1994. pp. 3-6

[10] Vladimirov VS. Generalized Functions in Mathematical Physics. Moscow: Mir; 1978. p. 282

[11] Alexeyeva LA, Kurmanov EB. Fundamental and generalized solutions of the two-component medium M. Bio 1. The Fourier transform of fundamental solutions and their regularization. Mathematical Journal. 2017;**17**(2):13-30

[12] Alexeyeva LA, Kurmanov EB. Fourier transform of fundamental solutions for the motion equations of two-component Biot's media. AIP Conference Proceedings. 2017;**1880**:1. DOI: 10.1063/1.5000675

Chapter 3

Boundary Integral Equations of no Stationary Boundary Value Problems for the Klein-Gordon Equation

Bayegizova Aigulim and Dadayeva Assiyat

Abstract

The non-stationary boundary value problems for Klein-Gordon equation with Dirichlet or Neumann conditions on the boundary of the domain of definition are considered; a uniqueness of boundary value problems is proved. Based on the generalized functions method, boundary integral equations method is developed to solve the posed problems in strengths of shock waves. Dynamic analogs of Green's formulas for solutions in the space of generalized functions are obtained and their regular integral representations are constructed in 2D and 3D over space cases. The singular boundary integral equations are obtained which resolve these tasks.

Keywords: boundary value problem, Klein-Gordon equation, generalized functions method, boundary integral equations, shock waves

1. Introduction

Klein-Gordon equation is hyperbolic differential equation in partial derivatives of second order. As is known, a class of solutions of hyperbolic equations contains no differentiable functions that have discontinuous derivatives on characteristic surfaces. By these cause, the construction of solutions for such equations, using smoothness of their differentiability, is impossible. However, this type of solution describes the shock waves, a presence of which is typical for physical processes.

Fundamental solutions of hyperbolic equations are singular generalized functions, the features of which are concentrated on moving surfaces—wave fronts, which propagate at a certain speed (it is called *light* or *sound speed*). This sharply distinguishes them from the fundamental solutions of elliptic and parabolic equations that are singular at a point. Therefore, classical methods of constructing boundary integral equations for solving nonstationary boundary value problems (BVP) based on the methods of potential theory or Green's formulas are unsuitable. To solve nonstationary boundary value problems for hyperbolic equations based on these methods, the Laplace or Fourier transform is usually used, which leads hyperbolic equations to parameterized elliptic type equations in the spaces of transforms. They are solved on the basis of the direct or indirect method of boundary integral equations (BIE). To restore original solutions, various numerical procedures of the inverse Fourier transform, Laplace, and others are used. The class of such problems of mathematical physics began to be considered as early as the 1970s of the last century, which was connected with the advent of computing technology, but so far, the number of works in this direction is very limited. As is known, the inverse transformation procedures are unstable, which are typical for solutions of the Fredholm equations of the first type and require regularization of the corresponding equations to obtain reliable results. Therefore, the problem of developing constructive methods for solving initial-boundary value problems for hyperbolic equations and systems of equations of mathematical physics for studying of various wave processes in bodies and in continuous medium remains relevant to the present time.

An effective method for solving boundary value problems for hyperbolic equations and systems of mixed type is the method of generalized functions (MGF), which allows to move from solving boundary value problems to solving the corresponding differential equations in the space of generalized functions and build integral representations of generalized solutions of initial boundary value problems in the original space-time to investigate the processes accompanied by shock waves.

To solve the Cauchy problem for hyperbolic equations, this method was proposed by Vladimirov [1]. In Refs. [2–6], the MGF was developed to solve nonstationary and stationary boundary value problems of the theory of elasticity, thermoelasticity, and electrodynamics and initial-boundary value problems for hyperbolic equations systems which are typical to the mathematical physics [7].

In the present chapter, this method is used to solve initial-boundary value problems for the Klein-Gordon equation (KG-Eq), a hyperbolic equation of the theory of elementary particles of quantum mechanics [8]. Here, nonstationary boundary value problems for KG-Eq with Dirichlet or Neumann conditions on the boundary of the domain of definition are considered and the uniqueness of the stated boundary value problems taking into account of shock waves is proved. Based on the method of generalized functions, the boundary integral equations method (BIEM) was developed to solve the stated tasks. Dynamic analogs of Green's formulas for their solutions in the space of generalized functions are obtained and regular integral representations in the plane and three-dimensional cases are constructed. Solving singular boundary integral equations are obtained for solving the stated initial-boundary value problems.

2. Klein-Gordon equation: shock waves

Klein-Gordon equation is formulated as:

$$\Box_c u + q(x)u = f(x,t). \qquad (1)$$

Here, we denote the wave operator,

$$\Box_c = \Delta - c^{-2}\frac{\partial^2}{\partial t^2},$$

where $\Delta = \sum_{k=1}^{N} \frac{\partial^2}{\partial x_k^2}$ is Laplace operator, $x \in R^N$, $t \in [0, \infty)$, and scattering potential $q(x) \in L_1(R^N)$. It is a hyperbolic equation. Its characteristic equation has the next form:

$$\nu_t^2 - c^2 \sum_{j=1}^{N} \nu_j^2 = 0, \tag{2}$$

where $\nu(x,t) = (\nu_1, \dots, \nu_N, \nu_t)$ is the normal vector to the characteristic surface F in R^{N+1}, $x \in R^N$, $(x,t) \in R^{N+1}$. It corresponds to the cone of characteristic normals —the light cone at $\nu_t < 0$. The solution of equation of $F(1)$ and its derivatives can be discontinuous. In R^N, characteristic surface F corresponds to wave front F_t (section of F at fixed t), which moves with speed c:

$$c = -\nu_t / \|\nu\|_N, \quad \|\nu\|_N = \sqrt{\nu_j \nu_j} \tag{3}$$

(here and further throughout in order to reduce the record on repeated indexes in the product, the summation from 1 to N is carried out, which is similar to tensor convolution). Such solutions (1) are called *shock waves*.

If solution of (1) is continuous:

$$[u(x,t)]_{F_t} = 0, \tag{4}$$

then wave fronts of Hadamard's conditions of continuity are satisfied under jumps:

$$[\dot{u} n_j + c u_{,j}]_{F_t} = 0, \quad j = \overline{1,N}; \tag{5}$$

$$[\dot{u} + c n_j u_{,j}]_{F_t} = 0, \tag{6}$$

for $x \in F_t$, where $n(x,t)$ is wave vector. It is a unit vector, normal to wave front F_t and directed forward its propagation. It is obvious that,

$$n_i = \nu_i / \|\nu\|_N, \quad i = \overline{1,N}; \tag{7}$$

Hereinafter, for abbreviation of the record, a symbol after comma defines the corresponding partial derivative: $u_{,j} = \frac{\partial u}{\partial x_j}$. The condition (5) is a consequence of the continuity condition (4) and ensures continuity on the F_t of tangent derivatives of u. The condition (5) is the law of conservation of momentum at the shock wave fronts. If before the wave front $u \equiv 0$, then at the wave front:

$$(grad\, u, n) = -c^{-1}\dot{u}, \quad x \in F_t.$$

To study the solutions of the KG-Eq, it is convenient to use the apparatus of the theory of generalized functions, which allows one to investigate shock waves, as well as singular solutions from the class of generalized functions typical for mathematical physics problems.

For this purpose, let us consider KG-Eq (1) on the space of generalized functions $D'(R^{N+1})$, which is the space of linear continuous functionals on the space of basic functions $D(R^{N+1})$, which are finite infinitely differentiable functions [1]. Further, the usual locally integrable function (*regular*) $f(x,t)$ will be marked from the top $\hat{f}(x,t)$, by considering it as a generalized.

If u (x,t) is regular differentiable and has a finite discontinuity on F, then in $D'(R^{N+1})$, as is known [1], its partial derivative is equal to the following:

$$\hat{u}_{,j} = u_{,j} + [u]_F \nu_j \delta_F, \tag{8}$$

where the first term on the right is the classical derivative of x_j, $\delta_F(x,t)$ is a simple layer on F, a singular generalized function [1], $\|\boldsymbol{\nu}\| = 1$. Using (8), it is possible to determine the second derivatives sequentially.

Definition. A solution $u(x, t)$ of Eq (1), which is continuous together with derivatives up to the second order almost everywhere, with the exception of a finite or countable number of discontinuity surfaces (wave fronts), on which the conditions (5) and (6) for jumps are satisfied, is called as classical solution.

Lemma 1. *If $u(x,t)$ is classic solution of* (1), *then $\hat{u}(x,t)$ is generalized solution of it.*

Proof. Taking into account these equations and (3), we get

$$(\Box_c + q(x))\hat{u} = f(x,t) + \left\{c^{-1}[u,_t]_{F_t} + \left[n_j u,_j\right]_{F_t}\right\}\|\nu\|_N \delta_F + c^{-1}\partial_t\left\{\|\nu\|_N [u]_{F_t}\delta_F)\right\}$$
$$+ \partial_j\left\{\|\nu\|_N [u]_{F_t} n_j \delta_F\right\}$$

By virtue of (4) and (6), the densities of simple and double layers here are equal to zero on the right, which were required to be proved.

From this lemma, it follows that the conditions on the fronts of shock waves are easy to obtain, considering the classical solutions of hyperbolic equations as generalized. It is enough to equate to zero the density of the corresponding independent singular generalized functions—analogs of simple, double, and other layers arising from the generalized differentiation of solutions. The determination of such conditions on the basis of classical methods is a very time-consuming procedure.

Let us put the energy density of the u-field E and the Lagrange L function:

$$E = 0.5\left(\dot{u}^2 + c^2\sum_{j=1}^{N} u,_j^2\right),$$

$$L = 0.5\left(\dot{u}^2 - c^2\sum_{j=1}^{N} u,_j^2\right).$$

Lemma 2. *If the classical solution of the KG-Eq.* (1), *then the following conditions for energy density jumps and Lagrange functions are satisfied at the shock wave fronts:*

$$[E]_{F_t} = -c\left[\dot{u}\frac{\partial u}{\partial n}\right]_{F_t},$$
$$[L(x,t)]_{F_t} = 0.$$

Proof: It is easy to show that for jumps, the equation is fulfilled:

$$[ab] = a^+[b] + b^-[a],$$

where the plus and minus signs indicate the limiting values of the functions a and b on the wave front from the side of the wave vector and opposite. Using this equality and the Hadamard's conditions (4) and (5), we get

$$\left[E + c\dot{u}\frac{\partial u}{\partial n}\right] = c^2\left[0.5(c^{-1}\dot{u}^2 + cu,_j u,_j) + \dot{u}\frac{\partial u}{\partial n}\right] = \ldots =$$
$$= 0.5c[\dot{u}]\left(\dot{u}^- + cu,_j^- n_j\right) + 0.5c^2[u,_j]\left(cu,_j^- + \dot{u}^- n_j\right) =$$
$$= 0.5c^3 u,_j^-\left[u,_j + c^{-1}n_j\dot{u}\right] + 0.5c\dot{u}^-\left[cn_j u,_j + \dot{u}\right] = 0$$

Here, n is the normal to the shock wave front in R^N. This implies the first formula of the lemma.

The condition for a jump in the energy density at the shock wave front can be obtained more easily by considering the corresponding energy equation in $D'\left(R^{N+1}\right)$, which we get by multiplying the Eq. (1) by $\dot{u}$. After simple transformations in the field of differentiability of solutions, we have the equation:

$$c^{-2}E,_t - \left(\dot{u}\, u,_j\right),_j + \dot{u}f = 0 \tag{9}$$

For shock waves in $D'\left(R^{N+1}\right)$, it has the form:

$$c^{-2}\dot{E} - \left(\dot{u}\, u,_j\right),_j + \dot{u}f = \left\{c^{-2}[E]\nu_t - \left[\dot{u}\, u,_j\right]\nu_j\right\}\delta_F =$$
$$= -\|\nu\|_N\left\{c^{-1}[E] + \left[\dot{u}\, u,_j\right]n_j\right\}\delta_{F_t}$$

For the right side to be turned to zero, it is necessary

$$[E] + c\left[\dot{u}\, u,_j\right]n_j = 0.$$

The latter coincides with the formula of Lemma 2.

Similarly, we can get the equation for L by multiplying Eq. (1) by u:

$$L + c^{-2}(u\dot{u}),_t - \left(uu,_j\right),_j + qu^2 - uf = 0 \tag{10}$$

Taking into account (3) and (4), for shock waves, we get

$$-[L] = \left[c^{-2}(u\dot{u}),_t - \left(uu,_j\right),_j\right] = u\left\{c^{-2}[\dot{u}]\nu_t - \left[u,_j\right]\nu_j\right\} = 0$$

It means that the Lagrange function is continuous at the shock wave fronts.

3. Statement of non-stationary BVP for Klein-Gordon equation: Energy conservation law

Let us construct the solution $u(x,\, t)$ of Eq. (1) on a set $S^- \in R^N$, bounded by surface S, by $t \geq 0$. Lets introduce next marks: $n(x)$ is vector of external normal to S; $D = \{S \times R^+\}$ is lateral surface of space-time cylinder $D^- = S^- \times R^+$, $R^+ = (0, +\infty)$; and the derivative of u on normal n at i, $\frac{\partial u}{\partial n} = u,_j n_j$.

Initial conditions: At $t = 0$ for $x \in S^-$:

$$u(x, 0) = u_0(x) \text{ for } x \in S^- + S \tag{11}$$

$$\dot{u}(x, 0) = v_0(x) \text{ for } x \in S^- \tag{12}$$

We consider two *boundary value problems* corresponding to the Dirichlet and Neumann conditions:

$$\text{(BVP I)} \quad u(x,t) = u_S(x,t) \text{ for } x \in S \tag{13}$$

$$\text{(BVP II)} \quad \frac{\partial u}{\partial n} = p(x,t) \text{ for } x \in S \tag{14}$$

At the shock wave fronts, the Hadamard conditions (5) and (6) on jumps are satisfied. Note that shock waves always occur if the condition of matching the initial and boundary data on the velocities is not satisfied

$$\dot{u}_S(x,0) = v_0(x), \quad x \in S, \tag{15}$$

which is typical for physical tasks. In this case, at the initial moment of time, a shock front is formed at the boundary S, which propagates with a velocity c in S^-. To construct continuously differentiable solutions, this condition is necessary. Here we will not enter it. Here, we not enter it and suppose that $u_0(x) \in C(S^- + S)$, $v_0(x) \in L_1(S^- + S)$, $p(x,t) \in L_1(D)$, and $u_S(x,t)$ a Holder's function on S: $\forall \beta, 0 < \beta \leq 1$, such that for $\forall x \in S, \ y \in S, \ t \geq 0$

$$|u_S(x,t) - u_S(y,t)| \leq const \|x - y\|^{\beta}, \tag{16}$$

here $L_1(\ldots)$ is the Lebeg's space of summable on the specified set of functions.

Let us mark as $D = \{S \times R^+\}$, the lateral surface of the space-time cylinder is $D^- = S^- \times R^+$, $R^+ = (0, +\infty)$.

Theorem 1. (*Energy conservation law*). *If $u(x,t)$ is classic solution of edge problem, then*

$$\int_{S^-} (E(x,t) - E_0(x))\, dV(x) + 0.5c^2 \int_{S^-} q(x)\left(u^2(x,t) - u_0^2(x)\right)\, dV(x)$$
$$= c^2 \int_0^t dt \int_S (\dot{u}_S(x,t) p(x,t)) dS(x) - c^2 \int_0^t dt \int_{S^-} f(x,t)\dot{u}(x,t)\, dV(x)$$

Proof. We integrate the energy Eq. (9) over a field with allowance for the partition of the field of integration by F_k wave fronts. Note that the first two terms can be considered as the divergence of the corresponding vector in space R^{N+1}, which is continuous in the regions between the fronts. Therefore, using the Ostrogradsky-Gauss theorem in R^{N+1}, we get

$$\int_{D^-} \left(c^{-2}E + \frac{1}{2}q(x)u^2\right)_{,t} dV(x,t) - \int_{D^-} (\dot{u}\, u_{,j})_{,j} dV(x,t)$$
$$+ \int_{D^-} \dot{u} f(x,t) dV(x,t) = \int_{D^-} \dot{u} f(x,t) dV(x,t) +$$
$$+ \int_{S^-} \left\{c^{-2}(E(x,t) - E_0(x)) + \frac{1}{2}q(x)\left(u^2(x,t) - u_0^2(x)\right)\right\} dV(x) -$$
$$- \int_0^t \int_S \left(\dot{u}\frac{\partial u}{\partial n}\right) dS(x) dt + \sum_{F_k} \int_{F_k} \left[c^{-2}E\nu_t - \frac{\partial u}{\partial \nu}\dot{u}\right]_{F_k} dF_k(x,t) = 0$$

Hereinafter, we denote $dV(x) = dx_1 \ldots dx_N$, $dV(x,t) = dV(x)dt$; $dF_k(x,t)$ is the differential of the surface area at the corresponding point of the wave front. By virtue of (3) and Lemma 2,

$$\left[c^{-2}E\nu_t - \dot{u}u_{,j}\nu_j\right]_{F_k} = -\|\nu\|_N c^{-1}\left[E + c\dot{u}\frac{\partial u}{\partial n}\right] = 0.$$

Therefore the last integral is zero. Taking into account the notation for the boundary functions, we get the formula of the theorem. From this theorem follows the Theorem 2.

Theorem 2. If $q(x) \geq 0$, *then the classic solution of first (second) BVP for Klein-Gordon equation is unique.*

Proof. Due to the linearity of the problem, it suffices to prove the uniqueness of the zero solution. For him, f = 0, the initial conditions and the corresponding boundary conditions are also zero. Then from Theorem 1, it follows:

$$\int_{S^-} \{E(x,t) + 0,5c^2 q(x) u^2(x,t)\} \ dV(x) = 0.$$

Since both terms are nonnegative, therefore, $E = 0$ and $u = 0$. The theorem is proved.

4. The dynamic analogue of Green's formula with constant scattering potential

Consider the case when scattering potential is constant:

$$q(x) = \pm m^2$$

To build the solution of BVP, we move to the space of generalized functions. To do this, we introduce the characteristic function of the solution domain

$$H_D^-(x,t) \equiv H_S^-(x) H(t),$$

where $H_S^-(x)$ is a characteristic function of set S^-, which is equal to 0.5 on its boundary S; and $H(t)$ is Heaviside's function, which is equal to 0.5 at $t = 0$. H_D^- is a characteristic function of space-time cylinder D^-. It is easy to show that:

$$\frac{\partial H_D^-}{\partial x_j} = -n_j \delta_S(x) H(t), \qquad \frac{\partial H_D^-}{\partial t} = -n_j H_S^-(x) \delta(t), \tag{17}$$

where $\delta(t)$ is singular Dirac's function.

To use the methods of the theory of generalized functions, we define the solution by zero outside the domain of the solution of the boundary value problem. For this, we put regular generalized functions:

$$\hat{u} = u(x,t) H_D^-(x,t), \quad \hat{f} = f(x,t) H_D^-(x,t), \tag{18}$$

where $u(x,t)$ is the classical solution of the BVP.

Consider the action of the KG-operator on $\hat{u}$. Since $[u]_S = -u$, and performing generalized differentiation using (17), we get

$$\Box_c \hat{u} \pm m^2 \hat{u} = -\frac{\partial u}{\partial n} \delta_S(x) H(t) - H(t) \left(u n_j \delta_S(x)\right)_{,j} - c^{-2} H_S^-(x) u_0(x) \dot{\delta}(t)$$
$$- c^{-2} H_S^-(x) \dot{u}_0(x) \delta(t) + \hat{f}(x,t), \tag{19}$$

where $\delta_S(x) H(t)$ is simple layer on lateral surface of a space-time cylinder $D = \{S \times R^+\}$.

Note that the densities of simple and double layers here are determined by the boundary conditions, some of which (depending on the boundary value problem) are known, and the given initial conditions.

The solution of Eq. (19) is convolution of the right part of the equation with its fundamental solution $\hat{U}(x,t)$, satisfying the conditions:

$$\Box_c \hat{U} \pm m^2 \hat{U} = \delta(x)\delta(t), \tag{20}$$

and *radiation conditions*:

$$\hat{U}(x,t) = 0 \ \text{ at } t < 0, \hat{U}(x,t) = 0 \ \text{ at } \|x\| > ct. \tag{21}$$

Let us call it the *Green's function* of Eq. (1).

The solution of (19) will be obtained in the form of the following convolution of right part of (19) and Green's function, which is equal to

$$\begin{aligned} \hat{u} = u(x,t)H_S^-(x)H(t) = & -\hat{U} * \frac{\partial u}{\partial n}\delta_S(x)H(t) - \left(\hat{U} * un_j\delta_S(x)H(t)\right)_{,j} - c^{-2}\left(\hat{U} \underset{x}{*} H_S^-(x)u_0(x)\right)_{,t} \\ & - c^{-2}\hat{U} \underset{x}{*} H_S^-(x)\dot{u}_0(x) + \hat{f}(x,t) * \hat{U} \end{aligned} \tag{22}$$

Here the symbol "*" means that convolution is taken only by x. Moreover, the solution is unique in the class of functions that allows convolution with U. Hence, it is easy to obtain a solution to the Cauchy problem (in the absence of S, $S^- = R^N$).

Consequence 1. The generalized solution of the Cauchy problem has the form:

$$\hat{u}(x,t) = -c^{-2}\hat{U} \underset{x}{*} \dot{u}_0 - c^{-2}\left(\hat{U} \underset{x}{*} u_0\right)_{,t} + \hat{f} * \hat{U} \tag{23}$$

Consequence 2. At zero initial data and $f = 0$, the generalized solution has the form:

$$\hat{u} = -\hat{U} * \frac{\partial u}{\partial n}\delta_S(x)H(t) - \hat{U}_{,j} * un_j\delta_S(x)H(t) \tag{24}$$

Formulas (22) and (24) express the solution of boundary value problems through the boundary values of the unknown function and its derivative along the normal to the boundary, i.e., they are similar to the Green formula for solutions of elliptic equations [9]. However, due to the singularities of the fundamental solutions of hyperbolic equations on the wave front, the form of which depends on the dimension of space, their integral representation gives divergent integrals containing derivatives of the fundamental solution. To construct regular integral representations, we introduce an antiderivative function:

$$\hat{W} = \hat{U} * \delta(x)H(t) = \hat{U} \underset{t}{*} H(t) \Rightarrow \partial_t \hat{W} = \hat{U} \tag{25}$$

and

$$\hat{H}(x,n,t) = \frac{\partial \hat{W}}{\partial x_j} n_j = \frac{\partial \hat{W}}{\partial n} \tag{26}$$

It is easy to see that $\hat{W}$и$\hat{H}$ are also solutions (1) at $\hat{f}(x,t) = H(t)\delta(x)$ and $\hat{f}(x,t) = H(t)\frac{\partial\delta(x,t)}{\partial n}$, respectively. The following theorem is true.

Theorem 3. *The generalized solution of boundary value problems has the form: (dynamic analogue of Green's formula)*

$$
\begin{aligned}
\hat{u} = & -\hat{U} * \frac{\partial u}{\partial n} \delta_S(x) H(t) - \hat{W}_{,j} * \dot{u} n_j(x) \delta_S(x) H(t) - \\
& - \hat{W}_{,j} \underset{x}{*} u_0(x) n_j(x) \delta_S(x) - c^{-2} \hat{U} \underset{x}{*} H_S^-(x) \dot{u}_0(x) - \\
& - c^{-2} \left(\hat{U} \underset{x}{*} H_S^-(x) u_0(x) \right)_{,t} + \hat{f} * \hat{U}
\end{aligned}
\tag{27}
$$

Proof. Let us consider the formula (22). It is easy to show, using the definition of the derivative of a generalized function and the continuity of u, that

$$(u n_j \delta_S(x) H(t))_{,t} = \dot{u}(x,t) n_j(x) \delta_S(x) H(t) + u(x,0) n_j(x) \delta_S(x) \delta(t)$$

Using this equality and the convolution differentiation property [1], we have

$$
\begin{aligned}
(\hat{U} * u n_j \delta_S(x) H(t))_{,j} &= (\hat{W}_{,t} * u n_j \delta_S(x) H(t))_{,j} \\
&= \hat{W}_{,j} * \dot{u}(x,t) n_j(x) \delta_S(x) H(t) + \hat{W}_{,j} * u(x,0) n_j(x) \delta_S(x) \delta(t)
\end{aligned}
$$

Since

$$\hat{W}_{,j} * u(x,0) n_j(x) \delta_S(x) \delta(t) = \hat{W}_{,j} \underset{x}{*} u_0(x) n_j(x) \delta_S(x),$$

putting these ratios in (22), we obtain the formula of the theorem.

From Theorem 3, it is consequent that the solution of the problem is entirely defined by initial data, boundary means of normal derivative of function $u(x,t)$, and its speed $\dot{u} = u_{,t} = \partial_t u$. By analog with representation of Laplace's equation solution, these formulas may be called *dynamical analog of Green's formula.*

Formula (27) of Theorem 3 allows at once to go to its integral writing without regularization of under integral functions on fronts.

Then let us consider representation of solution of edge problem for Klein-Gordon equations in spaces with dimensions N = 2,3, characterized for mathematical physics problems. To avoid complexity of formulas under building of integral representation of dynamical analog of Green's formula, let us consider consequently solutions of two BV problems:

1. Cauchy problem at $f(x,t) \neq 0$;

2. BVP at zero initial conditions and $f(x,t) = 0$.

By virtue of linearity of equations, solutions of BVPs may be obtained as a sum of solutions of these two problems with correction of boundary conditions for second problem with account of boundary meanings of Cauchy problem solutions. Solution of Cauchy problem for that equation has been early obtained by Vladimirov (see [9]). We get it here for the complete solution of the initial-boundary problem in the notation used here.

5. The generalized solution of the Cauchy problem for the KG-equation for N = 2

Let us consider the Cauchy problem for the KG equation of the below form:

$$\Box_c \hat{u} \pm m^2 \hat{u} = \hat{f}(x,t), x \in R^N, \quad t > 0, \tag{28}$$

where $\hat{f}(x,t)$ is a generalized function.
Let us introduce designations $r = \|y - x\|, S_t(x) = \{y \in S,\ r < ct\}$
$S_t^-(x) = \{y \in S^-,\ r < ct\}$ and $S_t(x) = \{y \in S,\ r < ct\}$, which we will use further.
In the flat case ($N = 2$), the Green's function of Eq. (28) is a regular generalized function of the form [9]:

$$\hat{U} = \frac{H(ct - \|x\|)}{2\pi} \frac{ch\left(m\sqrt{c^2t^2 - \|x\|^2}\right)}{\sqrt{c^2t^2 - \|x\|^2}} \tag{29}$$

with a weak singularity at the front $\|x\| = ct$:

$$\hat{U} \approx \frac{1}{2\pi\sqrt{c^2t^2 - \|x\|^2}} \qquad \text{by } \|x\| \to ct - 0 \tag{30}$$

Its carrier is a light cone: $\|x\| \le ct$.

Theorem 4. *If* $u_0(y) \in L_1(S^- + S)$, $\dot{u}_0(y) \in L_1(S^-)$, *then the solution of the Cauchy problem has the form:*

$$2\pi c^2 u(x,t)H(t) = \int_{S_t^-(x)} \frac{ch\left(m\sqrt{c^2t^2 - r^2}\right)}{\sqrt{c^2t^2 - r^2}} \dot{u}_0(y)dV(y) +$$

$$+\partial_t \int_{S_t^-(x)} \frac{ch\left(m\sqrt{c^2t^2 - r^2}\right)}{\sqrt{c^2t^2 - r^2}} u_0(y)dV(y) - \int_0^t d\tau \int_{S_\tau^-(x)} \frac{ch\left(m\sqrt{c^2\tau^2 - r^2}\right)}{\sqrt{c^2\tau^2 - r^2}} f(y,t-\tau)dV(y)$$

Proof. The integral notation of formula (23) leads to the formula of the theorem. All integrals are proper due to the regularity of integrands. The carrier of the kernel of integrals is a circle expanding over time with the center at the *point x*.

Note that if the initial conditions and the right-hand side of equation (1) (source) belong to the class of singular functions admitting convolution with the Green's function of the equation, to construct a solution to the Cauchy problem, use formulas (23) and (29).

Similarly, we construct a solution to the Cauchy problem in the case. The solution of the problem in this case allows analytic continuation. It can be obtained from the solution in Theorem 4 replacing ***m*** with im.

6. Generalized solution to the Cauchy problem for the KG equation for *N* = 3

For $N = 3$, the Green function (28) (for) is a singular generalized function of the form [9]:

$$4\pi\hat{U} = cH(t)\delta(c^2t^2 - r^2) - mcf_0(r,t) \tag{31}$$

where $r = \|x\|$, $H(t)\delta(c^2t^2 - r^2)$ is a simple layer on a light cone $r = ct$ [9].

The function f_0 *is defined by the expression:*

$$f_0(r,t) = \frac{H(ct-r)J_1\left(m\sqrt{c^2t^2-r^2}\right)}{\sqrt{c^2t^2-r^2}} \tag{32}$$

$J_1(\ldots)$ is *Bessel function. Because* [10],

$$J_1(z) \sim 0,5\,z \qquad \text{when} \quad z \to 0 \tag{33}$$

at the front $r = ct$, the second term has a finite jump:

$$[f_0(r, r/c)] = -\frac{m}{2} \tag{34}$$

Theorem 5. *The solution of the Cauchy problem for the KG-Eq.* (28) *for N = 3 has the form:*

$$4\pi c u(x,t) = (2ct)^{-1} \int_{r=ct} \dot{u}_0(y)\,dS(y) - m \int_{S_t^-(x)} \frac{J_1\left(m\sqrt{c^2t^2-r^2}\right)}{\sqrt{c^2t^2-r^2}} \dot{u}_0(y)\,dV(y) +$$

$$+\frac{1}{2ct^2}\int_{r=ct} u_0(y)dS(y) + \frac{1}{2ct}\partial_t \int_{r=ct} u_0(y)dS(y) - m\partial_t\left\{\int_{S_t^-(x)} \frac{J_1\left(m\sqrt{c^2t^2-r^2}\right)}{\sqrt{c^2t^2-r^2}} u_0(y)\,dV(y)\right\} -$$

$$-mc^2\int_0^t d\tau \int_{S_\tau^-(x)} \frac{f(y,t-\tau)J_1\left(m\sqrt{c^2\tau^2-r^2}\right)}{\sqrt{c^2\tau^2-r^2}} dV(y) + \frac{c}{2}\int_{S_t^-(x)} r^{-1} f(y, t-r/c)\,dV(y)$$

Proof. It follows from the representation of a generalized solution for the Cauchy problem taking into account the form of the fundamental solution (30). The solution of the Cauchy problem for Eq. (28) in the case $q(x) = -m^2$ also allows analytic continuation by replacing m with im. It has the form:

$$4\pi c u(x,t) = \frac{1}{2t}\int_{r=ct} \dot{u}_0(y)\,dS(y) - m \int_{S_t^-(x)} \frac{I_1\left(m\sqrt{c^2t^2-r^2}\right)}{\sqrt{c^2t^2-r^2}} \dot{u}_0(y)\,dV(y) +$$

$$+\frac{1}{2ct^2}\int_{r=ct} u_0(y)dS(y) + \frac{1}{2ct}\partial_t \int_{r=ct} u_0(y)dS(y) -$$

$$-m\partial_t\left\{\int_{S_t^-(x)} \frac{I_1\left(m\sqrt{c^2t^2-r^2}\right)}{\sqrt{c^2t^2-r^2}} u_0(y)\,dV(y)\right\} +$$

$$+\frac{c}{2}\int_{S_t^-(x)} r^{-1} f\left(y, t-\frac{r}{c}\right)dV(y) - mc^2\int_0^t d\tau \int_{S_\tau^-(x)} \frac{f(y,t-\tau)I_1\left(m\sqrt{c^2\tau^2-r^2}\right)}{\sqrt{c^2\tau^2-r^2}} dV(y).$$

If the initial functions and the right-hand side of Eq. (1) belong to the class of singular functions admitting convolution with the Green function of Eq. (28), to construct the solution, one should use the formula in ultraprecise form (23). We construct solutions to initial-boundary value problems.

7. Singular boundary integral equations of plane boundary value problems

Let us consider the solutions of the posed boundary value problems in the case $N = 2$. For the integral representation of the dynamic analogue of the Green formula, we also calculate for the Green function (29):

$$\hat{W} = \frac{1}{2\pi} d_0(r,t) * H(t) = \frac{1}{2\pi c} d_1(r,t), \tag{35}$$

$$\hat{H}(x,t,n) = \frac{1}{2\pi c} d_2(r,t) \frac{\partial r}{\partial n}, \quad r = \|x\|, \tag{36}$$

where

$$d_0(r,t) = H(ct-r) \frac{ch\left(m\sqrt{c^2t^2 - r^2}\right)}{\sqrt{c^2t^2 - r^2}},$$

$$d_1(r,t) = H(ct-r) \int_0^{\sqrt{c^2t^2-r^2}} \frac{ch(mz)}{\sqrt{z^2 + r^2}} dz,$$

$$d_2(r,t) = -H(ct-r)\left\{\frac{1}{r} - \frac{ch\left(m\sqrt{c^2t^2 - r^2}\right)}{ct}\right\} - H(ct-r) r \int_0^{\sqrt{c^2t^2-r^2}} \frac{ch(mz)}{\left(\sqrt{z^2 + r^2}\right)^3} dz.$$

Consider the values of these functions at the front $r = ct$, $t > 0$. From (29), (27) follows that

$$W\big|_{r=ct} = 0, \quad H\big|_{r=ct} = 0, \tag{37}$$

Consequently, unlike from U, W and H are continuous at the front. When $r \to 0$, we have an asymptotic representation:

$$U = \frac{ch(mct)}{2\pi ct} + O(r), \quad H = -\frac{1}{2\pi cr} \frac{\partial r}{\partial n} + O(1) \tag{38}$$

Now we turn to the integral notation of the dynamic analogue of Green's formula for $N = 2$.

Theorem 6. *The solution of the initial-boundary value problem for the KG-equation in the flat case is representable: for* $x \notin S$ *in form of*

$$2\pi u(x,t)H(t) = \int_{S_t(x)} H(ct-r) dS(y) \int_{r/c}^{t} \left\{ d_2(r,\tau) \frac{\partial r}{c\partial n(y)} \dot{u}(y,t-\tau) \right\} d\tau -$$

$$- \int_{S_t(x)} H(ct-r) dS(y) \int_{r/c}^{t} d_0(r,\tau) p(y,t-\tau) d\tau, \quad r = \|y - x\|;$$

for $x \in S$ *in form of*

$$\pi c u(x,t) H(t) = V.P. \int_{S_t(x)} H(ct-r) \frac{\partial r}{\partial n(y)} dS(y) \int_{r/c}^{t} d_2(r,\tau) \dot{u}(y,t-\tau) d\tau -$$

$$-c \int_{S_t(x)} H(ct-r) dS(y) \int_{r/c}^{t} d_0(r,\tau) p(y,t-\tau) d\tau.$$

Proof. In the flat case, the formula of Theorem 3, taking into account the carriers of the cores, can be written in the integral form:

$$2\pi \hat{u}(x,t) = \int_{S_t(x)} H(ct-r) dS(y) \int_{r/c}^{t} H(x,y,\tau,n(y)) \dot{u}(y,t-\tau) d\tau -$$

$$-H(t) \int_{S_t(x)} H(ct-r) dS(y) \int_{r/c}^{t} U(x,y,\tau) \frac{\partial u(y,t-\tau)}{\partial n(y)} d\tau.$$

Substituting the form of the cores (29), (6.2), we obtain the first formula of the theorem. For $x \notin S$, all integrals are convergent, because $r \neq 0$, and U has an integrable singularity at the front (30). Let us prove the second formula for $x^* \in S$.

We write a dynamic analogue of Green's formula for a region with a puncture ε-vicinity at point x^*. We denote $\Gamma_\varepsilon^-(x) = \{y \in S^- : \|y-x^*\| = \varepsilon\}$, $O_\varepsilon(x) = \{y \in S : \|y-x^*\| < \varepsilon\}$, $\Omega_\varepsilon(x^*) = \{S_t(x^*) - O_\varepsilon(x^*)\} \cup \Gamma_\varepsilon^-(x^*)$. Because the x^* outside the area bounded by a compound path $\Omega_\varepsilon(x^*)$:

$$\Sigma_{1\varepsilon} + \Sigma_{2\varepsilon} == \int_{\Omega_\varepsilon(x^*)} H(ct-r) \frac{\partial r}{\partial n(y)} dS(y) \int_{r^*/c}^{t} d_2(r,\tau) \dot{u}(y,t-\tau) \frac{d\tau}{c} -$$

$$- \int_{\Omega_\varepsilon(x^*)} H(ct-r) dS(y) \int_{r/c}^{t} d_0(r,\tau) p(y,t-\tau) d\tau = 0$$

In this equality, we pass to the limit for $\varepsilon \to +0$. For $\Gamma_\varepsilon^-(x^*)$ $r = \varepsilon$, arc length differential in polar coordinates: $dS(y) = \varepsilon d\varphi$, therefore

$$\lim_{\varepsilon \to 0} \int_{\Gamma_\varepsilon^-(x^*)} H(ct-\varepsilon) dS(y) \int_{\varepsilon/c}^{t} d_0(\varepsilon,\tau) p(y,t-\tau) d\tau =$$

$$= \lim_{\varepsilon \to 0} \varepsilon \pi \int_{\varepsilon/c}^{t} \left\{ \frac{ch(mc\tau)}{c\tau} p(x^*,t-\tau) \right\} d\tau = \pi c \lim_{\varepsilon \to 0} \varepsilon \int_{1}^{ct/\varepsilon} \left\{ \frac{ch(m\varepsilon\tau)}{\tau} p\left(x^*, t - \frac{\varepsilon}{c}\tau\right) \right\} d\tau = 0$$

We have the last equality due to the inequality:

$$\left| \int_{1}^{ct/\varepsilon} \left\{ \frac{ch(m\varepsilon\tau)}{\tau} p\left(x^*, t - \frac{\varepsilon}{c}\tau\right) \right\} d\tau \right| \leq ch(mct) \int_{0}^{\infty} |p(x^*,\tau)| d\tau$$

Consider the limit of the first integral:

$$\lim_{\varepsilon\to 0}\Sigma_{1\varepsilon}=\lim_{\varepsilon\to 0}\int_{\Omega_\varepsilon(x^*)} H(ct-r)\frac{\partial r}{\partial n(y)}dS(y)\int_{r^*/c}^{t} d_2(r,\tau)\dot{u}(y,t-\tau)\frac{d\tau}{c}=$$

$$=V.P.\int_{S_t(x^*)} H(ct-r)\frac{\partial r}{\partial n(y)}dS(y)\int_{r^*/c}^{t} d_2(r,\tau)\dot{u}(y,t-\tau)\frac{d\tau}{c}- \quad (39)$$

$$-\lim_{\varepsilon\to 0}\int_{\Gamma_\varepsilon^-(x^*)} H(ct-r)\frac{\partial r}{\partial n(y)}dS(y)\int_{r/c}^{t} d_2(r,\tau)\dot{u}(y,t-\tau)\frac{d\tau}{c}=I_S+\lim_{\varepsilon\to 0}I_\varepsilon$$

We calculate the last limit on the right side.

Since on $\Gamma_\varepsilon^-(x^*)$: $\dfrac{\partial r}{\partial n}=-1\Rightarrow$

$$H=\frac{1}{2\pi c\varepsilon}+O(1),$$

Therefore

$$\lim_{\varepsilon\to 0}I_\varepsilon=-\lim_{\varepsilon\to 0}\int_{\Gamma_\varepsilon^-(x^*)}\frac{\partial r}{\partial n(y)}dS(y)\int_{\varepsilon/c}^{t} d_2(\varepsilon,\tau)\dot{u}(x^*,t-\tau)\frac{d\tau}{c}=$$

$$=\lim_{\varepsilon\to 0}\pi\varepsilon\int_{\varepsilon/c}^{t}\frac{\dot{u}(x^*,t-\tau)}{\varepsilon}d\tau=\pi\lim_{\varepsilon\to 0}\int_0^t\dot{u}(x^*,t-\tau)\,d\tau=$$

$$=\pi\left(u(x^*,0)-u(x^*,t)\right)=-\pi u(x^*,t)$$

Adding, taking into account the last equality, we obtain the second formula of the theorem. The theorem has been proved.

In the case of the first boundary value problem, the left side of the equation of Theorem 6.1 and the first integral on the right are known, determined by the boundary conditions. Solving it, we determine the normal derivative of the desired function on the boundary, after which the formula of the theorem allows us to calculate the solution at any point in the domain of definition. In the case of the second boundary-value problem, we have a BIE to determine the unknown boundary values of the unknown function *u* from the boundary values of its normal derivative. Solving it, we determine its values at the border, after which we determine the solution.

8. Dynamic analog of Green's formula for solutions of the KG-equation (N = 3)

To construct a dynamic analogue of Green's formula in integral form, we define $\hat{W}$ и $\hat{H}$. By computing formulas (25)and (27), we obtain

$$4\pi c^2\hat{W}=\frac{H(ct-r)}{r}-m\,f_1(r,t),\qquad r=\|x\| \quad (40)$$

$$\hat{H}(x,n,t) = -\frac{(x,n)}{2r}\left(\frac{\delta(ct-r)}{r} + \frac{H(ct-r)}{r^2} + 2mf_2(r,t)\right),$$

where $f_1(r,t) = H(ct-r)\int\limits_0^{\sqrt{c^2t^2-r^2}} \frac{J_1\left(\frac{m}{c}z\right)}{\sqrt{z^2+r^2}}dz,$

$$f_2(r,t) = rH(ct-r)\int\limits_0^{\sqrt{c^2t^2-r^2}} \frac{J_1\left(\frac{m}{c}z\right)}{\left(\sqrt{z^2+r^2}\right)^3}dz + \frac{rH(ct-r)}{ct\sqrt{c^2t^2-r^2}}J_1\left(\frac{m}{c}\sqrt{c^2t^2-r^2}\right) \tag{41}$$

The values of these functions at the front $r = ct$:

$$f_1(r, r/c) = 0, \quad f_2(r, r/c) = \frac{m}{2c}$$

Consequently, W is continuous at the wave front; the last term in the representation H has a discontinuity of the first kind at the front. When $r \to 0$, $t > 0$, the following asymptotics are true:

$$\hat{W}(x,t) = \frac{H(t)}{8\pi c^2 r} + O(1) \tag{42}$$

$$\hat{H}(x,n,t) = \frac{(e_x,n)}{8\pi c^2 r^2}H(t) + O(r^{-1}) \tag{43}$$

Let us impose the notation for the shift functions:

$$U(x,y,t) = \hat{U}(x-y,t),\ W(x,y,t) = \hat{W}(x-y,t),$$
$$H(x,y,t,n) = \hat{H}(x-y,t,n)$$

Theorem 7. *The generalized solution of boundary value problems for a homogeneous KG-equation satisfying zero initial conditions* $(u(x,0) = 0, \dot{u}(x,0) = 0)$, *is representable in the form:*

$$4\pi\hat{u}(x,t) = -H(t)\int\limits_{S_t(x)} \frac{1}{r}\frac{\partial u\left(y, t-\frac{r}{c}\right)}{\partial n(y)}dS(y) - c^{-1}H(t)\int\limits_{S_t(x)} \dot{u}\left(y, t-\frac{r}{c}\right)\frac{(y-x, n(y))}{r^2}dS(y) +$$

$$+mc^{-1}H(t)\int\limits_{S_t^-(x)} f_0(r, r/c)\frac{\partial u(y, t-\tau)}{\partial n(y)}dV(y) + mH(t)\int\limits_0^t d\tau \int\limits_{S_\tau(x)} f_2(r,\tau)\dot{u}(y, t-\tau)\ dS(y) +$$

$$+H(t)\int\limits_{S_t(x)} u\left(y, t-\frac{r}{c}\right)\frac{(y-x, n(y))}{r^3}dS(y),$$

For $x \in S$, *the last integral is singular, taken in the sense of the principal value.*

Proof. Using the conditions of Theorem and (30), we write in this case a dynamic analogue of Green's formula (22). We compute convolution sequentially:

$$4\pi\hat{U}(x,t)*\frac{\partial u}{\partial n}\delta_S(x)H(t)=\int\limits_{S_t(x)}\frac{1}{r}\frac{\partial u(y,t-r/c)}{\partial n(y)}dS(y)-mc\int\limits_0^t d\tau\int\limits_{S_\tau(x)}f_0(r,\tau)\frac{\partial u(y,t-\tau)}{\partial n(y)}dS(y)=$$

$$\int\limits_{S_t(x)}\frac{1}{r}\frac{\partial u(y,t-r/c)}{\partial n(y)}dS(y)-m\int\limits_{S_t^-(x)}f_0(r,r/c)\frac{\partial u(y,t-\tau)}{\partial n(y)}dV(y);$$

$$-4\pi\hat{W}_{,j}*\dot{u}(x,t)n_j(x)\delta_S(x)H(t)=-\int\limits_{S_t(x)}\dot{u}(y,t-r/c)\frac{(y-x,n(y))}{cr^2}dS(y)-$$

$$-\int\limits_0^t d\tau\int\limits_{S_\tau(x)}\frac{(y-x,n(y))}{r^3}\dot{u}(y,t-\tau)\,dS(y)+\int\limits_0^t d\tau\int\limits_{S_t(x)}\frac{(y-x,n(y))}{c^2r}f_2(r,\tau)\,\dot{u}(y,t-\tau)dS(y).$$

When $r=\|y-x\|\neq 0$, you can change the order of integration, therefore:

$$\int\limits_0^t d\tau\int\limits_{S_\tau(x)}\frac{(y-x,n(y))}{r^3}\dot{u}(y,t-\tau)\,dS(y)=-\int\limits_{S_t(x)}u(y,t-r/c)\frac{(y-x,n(y))}{r^3}dS(y)$$

Summing up, we obtain the formula of the theorem.

Note that for points $x\in S$ when $t>0$, all cores have no singularities and the integrals on the right exist and define functions that are regular on a given set. Since regular functions are on left and right and they are equal on this set as generalized, by virtue of the du Bois-Reymond lemma [1], they are equal in the usual sense, like numerical functions.

Let us assume $x^*\in S$. We write a dynamic analogue of Green's formula for a region with a puncture $\varepsilon-$vicinity point $x^*\in S,\varepsilon<ct$:

$$\begin{aligned}0=&-\int\limits_{S_t(x)-O_\varepsilon(x)}\frac{1}{r}\frac{\partial u\left(y,t-\frac{r}{c}\right)}{\partial n(y)}dS(y)-\int\limits_{S_t(x)-O_\varepsilon(x)}\dot{u}\left(y,t-\frac{r}{c}\right)\frac{(y-x,n(y))}{cr^2}dS(y)+\\&+mc\int\limits_0^t d\tau\int\limits_{S_\tau(x)-O_\varepsilon(x)}f_0(r,\tau)\frac{(y-x,n(y))}{r}\frac{\partial u(y,t-\tau)}{\partial n(y)}dS(y)+\int\limits_0^t d\tau\int\limits_{S_\tau(x)-O_\varepsilon(x)}f_2(r,\tau)\dot{u}(y,t-\tau)dS(y)-\\&-\int\limits_{\Gamma_\varepsilon^-(x,t)}\frac{1}{r}\frac{\partial u(y,t-r/c)}{\partial n(y)}dS(y)-\int\limits_{\Gamma_\varepsilon^-(x,t)}\dot{u}(y,t-r/c)\frac{(y-x,n(y))}{cr^2}dS(y)+\\&+mc\int\limits_0^t d\tau\int\limits_{\Gamma_\varepsilon^-(x,\tau)}f_0(r,\tau)\frac{(y-x,n(y))}{r}\frac{\partial u(y,t-\tau)}{\partial n(y)}dS(y)+\int\limits_0^t d\tau\int\limits_{\Gamma_\varepsilon^-(x,\tau)}f_2(r,\tau)\dot{u}(y,t-\tau)\,dS(y)+\\&+\int\limits_{S_t(x)-O_\varepsilon(x)}u(y,t-r/c)\frac{(y-x,n(y))}{r^3}dS(y)+\int\limits_{\Gamma_\varepsilon^-(x,t)}u(y,t-r/c)\frac{(y-x,n(y))}{r^3}dS(y),\end{aligned}\quad(44)$$

Now let us move on to the limit $\varepsilon\to 0$. In the first integral, the integrand has a weak integrable singularity when r = 0. In the second integral, it does not have a singularity when r = 0. By virtue of this, the integrals over Γ_ε^- from these functions in the third and fourth term tend to zero. It is obvious that

$$\lim_{\varepsilon\to 0}\int\limits_{S_t(x)-O_\varepsilon(x)}u(y,t-r/c)\frac{(y-x,n(y))}{2r^3}dS(y)=V.P.\int\limits_{S_t(x)}u(y,t-r/c)\frac{(y-x^*,n(y))}{2r^3}dS(y)$$

Moreover, the main value of the integral exists, because the integrand has a singularity of order $1/r^2$ on a two-dimensional surface S, function u is continuous on S, and the characteristic $(y-x^*, n(x^*))$ antisymmetric in opposite relative x^* points.

Let us consider the last limit. For $\Gamma_\varepsilon^-(x)$, we have $\|y-x\|=\varepsilon$, $n(y)=(x-y)/\varepsilon$, therefore

$$\lim_{\varepsilon\to 0}\left\{\int_{\Gamma_\varepsilon^-(x^*,t)} u(y,t-r/c)\frac{(y-x^*,n(y))}{r^3}dS(y)\right\}=$$

$$=\lim_{\varepsilon\to 0}\left\{\int_{\Gamma_\varepsilon^-(x^*,t)}\left(u(y,t-r/c)-u(x^*,t)\right)\frac{(-n(y),n(y))}{\varepsilon^2}dS(y)\right\}-u(x^*,t)\lim_{\varepsilon\to 0}\left\{\int_{\Gamma_\varepsilon^-(x^*,t)}\frac{dS(y)}{\varepsilon^2}\right\}=-2\pi u\left(x^*,t\right)$$

Passing in (44) to the limit in $\varepsilon\to 0$ and transferring the last term to the right side, we obtain the formula of the theorem. The theorem has been proved.

Formula ($x\in S$) gives a singular boundary integral equation for solving the second initial-boundary value problem. For the first boundary value problem, the unknown normal derivative falls under the sign of the surface integral with a weakly polar core. The remaining terms are known.

9. Conclusion

Note that the constructed delayed singular BIE have a nonclassical type; since in addition to the boundary values $u,_t$ of the function and its normal derivative, the BIE includes a velocity that is unknown for the Dirichlet problem and is known for the Neumann problem. In addition, the integration region at the boundary depends on time, which also distinguishes these equations from the SEI for elliptic and parabolic problems. Solving the Dirichlet problem on the basis of the method of successive approximations, like elliptic problems, is impossible, since it requires the determination of boundary values of velocity. However, differentiation of generalized solutions on the boundary leads to hypersingular relations. This is a new class of BIE in delayed potentials, which requires a special study by the methods of functional analysis. However, to solve the resolving singular BIE that solve the boundary value problems, numerical boundary element method can be used.

Acknowledgements

This work was financially supported by the Ministry of Education and Science of the Republic of Kazakhstan (grant AP05132272).

Author details

Bayegizova Aigulim[1*] and Dadayeva Assiyat[2]

1 L.N. Gumilyov Eurasian National University, Nur-Sultan, Kazakhstan

2 Institute of Mathematics and Mathematical Modelling, Almaty, Kazakhstan

*Address all correspondence to: baegiz_a@mail.ru

References

[1] Vladimirov VS. Generalized Functions in Mathematical Physics. Moscow: Mir; 1978. p. 282

[2] Alekseeva LA, Dildabaev SA, Zhanbyrbaev AB, Zakiryanova GK. Boundary integral equation method in two & three dimensional problems of elastodynamics. International Journal of Computational Mechanics. 1996;**18**(2): 147-157

[3] Alekseeva LA. Boundary element method of boundary value problems of elastodynamics by stationary running loads. International Journal of Engineering Analysis with Boundary Element. 1998;(11):37-44

[4] Alekseeva LA. Singular boundary integral equations of boundary value problems of elastodynamics in the case of subsonic traveling loads. Differential Equations. 2010;**46**(4):512-519

[5] Alexeyeva LA. Generalized solutions of non-stationary boundary value problems for Maxwell's equations. Journal of Computational Mathematics and Mathematical Physics. 2002; **42**(1):76-88

[6] Alexeyeva LA, Kupesova BN. The method of generalized functions in boundary value problems of coupled thermoelastodynamics. Applied Mathematics and Mechanics. 2001; **65**(2):334-345

[7] Alexeyeva LA, Zakiryanova GK. Generalized solutions of initial-boundary value problems for second-order hyperbolic systems. Journal of Computational Mathematics and Mathematical Physics. 2011;**51**(7): 1280-1293

[8] Bogolyubov NN, Logunov AA, Oksak AI, Todorov IT. General Principles of Quantum Field Theory. M: Science; 1987

[9] Polyanin AD. Handbook of Linear Equations of Mathematical Physics. M: FIZMATLIT; 2001. 576 p. ISBN: 5-9221-0093-9

[10] Abramovits M, Stegana I. Handbook of Special Functions. M: Science; 1979

Chapter 4

Singular Boundary Integral Equations of Boundary Value Problems for Hyperbolic Equations of Mathematical Physics

Lyudmila A. Alexeyeva and Gulmira K. Zakiryanova

Abstract

The method of boundary integral equations is developed for solving the nonstationary boundary value problems (BVP) for strictly hyperbolic systems of second-order equations, which are characteristic for description of anisotropic media dynamics. The generalized functions method is used for the construction of their solutions in spaces of generalized vector functions of different dimensions. The Green tensors of these systems and new fundamental tensors, based on it, are obtained to construct the dynamic analogues of Gauss, Kirchhoff, and Green formulas. The generalized solution of BVP has been constructed, including shock waves. Using the properties of integrals kernels, the singular boundary integral equations are constructed which resolve BVP. The uniqueness of BVP solution has been proved.

Keywords: hyperbolic equations, generalized solution, Green tensor, boundary value problem, generalized function method

1. Introduction

Investigation of continuous medium dynamics in areas with difficult geometry with various boundary conditions and perturbations acting on the medium leads to boundary value problems for systems of hyperbolic and mixed types. An effective method to solve such problems is the boundary integral equation method (BIEM), which reduces the original differential problem in a domain to a system of boundary integral equations (BIEs) on its boundary. This allows to lower dimension of the soluble equations, to increase stability of numerical procedures of the solution construction, etc. Note that for hyperbolic systems, BIEM is not sufficiently developed, while for solving boundary value problems (BVPs) for elliptic and parabolic equations and systems, this method is well developed and underlies the proof of their correctness. It is connection with the singularity of solutions to wave equations, which involve characteristic surfaces, i.e., wavefronts, where the solutions and their derivatives can have jump discontinuities. As a result, the fundamental solutions on wavefronts are essentially singular, and the standard methods for constructing BIEs typical for elliptic and parabolic equations cannot be used. Therefore, for the development of the BIEM for hyperbolic equations, the theory of generalized functions [1, 2] is used. At present, BIEM are applied very extensively to solve engineering problems.

Here, the second-order strictly hyperbolic systems in spaces of any dimension are considered. The fundamental solutions of consider systems of equations are constructed and their properties are studied. It is shown that the class of fundamental solutions for our equations in spaces of odd dimensions is described by singular generalized functions with a surface support (e.g. for $R^3 \times t$, this is a single layer on a light cone). The constructed fundamental solutions of consider systems of equations are the kernels of BIEs. For systems of hyperbolic equations, the BIE method is developed. Here, the ideas for solving nonstationary BVPs for the wave equations in multidimensional space [3, 4] are used and the methods were elaborated for boundary value problems of dynamics of elastic bodies [5–8].

2. Generalized solutions and conditions on wave fronts

Consider the second-order system of hyperbolic equations with constant coefficients:

$$L_{ij}(\partial_x, \partial_t)u_j(x,t) + G_i(x,t) = 0, \quad (x,t) \in R^{N+1} \tag{1}$$

$$L_{ij}(\partial_x, \partial_t) = C_{ij}^{ml}\partial_m\partial_l - \delta_{ij}\partial_t^2, \quad i,j = \overline{1,M}, \quad m,l = \overline{1,N} \tag{2}$$

$$C_{ij}^{ml} = C_{ij}^{lm} = C_{ji}^{ml} = C_{ml}^{ij} \tag{3}$$

where $G_i \in L_2(R^{N+1})$ and δ_{ij} are Kronecker symbols; $\partial_x = (\partial_1, \dots \partial_N)$, $\partial_i = \partial/\partial x_i$, and $\partial_t = \partial/\partial t$ are Partial derivatives; and also we will use following notations $u_{i,j} = \partial_j u_i$ and $u_{i,t} = \partial_t u_i$.

The matrix C_{ij}^{ml}, whose indices may be permitted in accordance with above indicated symmetry properties (3), satisfies the following condition of strict hyperbolicity:

$$W(n,v) = C_{ij}^{ml} n_m n_l v^i v^j > 0 \quad \forall n \neq 0, \quad v \neq 0$$

Here everywhere like numbered indices indicate summation in specified limits of their change (so as in tensor convolutions).

By the virtue of positive definiteness W, the characteristic equation of the system (1)

$$det\left\{C_{ij}^{ml} n_m n_l - c^2\delta_{ij}\right\} = 0, \|n\| = 1 \tag{4}$$

has 2M valid roots (with the account of multiplicity):

$$c = \pm c_k(n) : 0 < c_k \leq c_{k+1}, k = \overline{1, M-1}$$

They are *sound velocities* of wave prorogations in physical media which are described by such equations. In a general case, they depend on a wave vector n.

It is known that the solutions of the hyperbolic equations can have characteristic surfaces on which the jumps of derivatives are observed [9]. To receive the conditions on jumps, it is convenient to use the theory of generalized functions.

Denote through $D'_M(R^{N+1})$ the space of generalized vector functions $\hat{f}(x,t) = \left(\hat{f}_1, \dots, \hat{f}_M\right)$ determined on the space $D_M(R^{N+1})$ of finite and indefinitely differentiable vector functions $\varphi(x,t) = (\varphi_1, \dots, \varphi_M)$. For regular $\hat{f}$, this linear function is presented in integral form:

$$\left(\hat{f}(x,t), \varphi(x,t)\right) = \int_{-\infty}^{\infty} d\tau \int_{R^N} f_i(x,\tau)\varphi_i(x,\tau)dV(x), \quad \forall\varphi \in D_M\left(R^{N+1}\right), \quad i = \overline{1,M}$$

$dV = dx_1 \dots dx_N$ (further, we shall say everywhere *generalized function* instead of *generalized vector function*).

Let $u(x,t)$ be the solution of Eq. (1) in R^{N+1}, continuous, twice differentiable almost everywhere, except for characteristic surface F which is motionless in R^{N+1} and mobile in R^N (wave front F_t). On surface, F_t derivatives can have jumps. The equation of F is Eq. (4). We denote $\nu = (n_1, \dots {}_{nN}, n_t) = (n, t_t), n = (n_1, \dots {}_{nN})$, where ν is a normal vector to the characteristic surface F in R^{N+1}, and n is unit wave wave vector in R^N directed in the direction of propagation F_t. It is assumed that the surface F is piecewise smooth with continuous normal on its smooth part.

Let us consider Eq. (1) in the space $D'_M\left(R^{N+1}\right)$ and its solutions in this space are named as *generalized solutions* of Eq. (1) (or *solutions in generalized sense)*.

The solution $u(x,t)$ is considered as a regular generalized function and we denote $\hat{u}(x,t) = u(x,t)$, accordingly $\hat{G}(x,t) = G(x,t)$. Let $\hat{u}(x,t)$ be the solution of Eq. (1) in $D'_M\left(R^{N+1}\right)$.

Theorem 2.1. *If $\hat{u}(x,t)$ is the generalized solution of Eq. (1), then there are next conditions on the jumps of its components and derivatives*:

$$[u_i(x,t)]_{F_t} = 0 \tag{5}$$

$$\left[\sigma_i^m n_m - c u_{i,t}\right]_{F_t} = 0 \tag{6}$$

where $\sigma_i^m = C_{ij}^{ml} u_{j,l}$ and the velocity c of a wave front F_t coincides with one of c_k.

Proof. By the account of differentiation of regular generalized function rules [2], we receive:

$$\begin{aligned} L_{ij}(\partial_x, \partial_t)\hat{u}_j(x,t) + \hat{G}_i(x,t) &= \left[\sigma_i^m \nu_m - \nu_t u_{i,t}\right]_F \delta_F(x,t) + \\ &+ C_{ij}^{ml}\partial_m\left([u_j]_F \nu_l \delta_F(x,t)\right) - \left([u_i]_F \nu_t \delta_F(x,t)\right)_{,t} \end{aligned} \tag{7}$$

Here, $\alpha(x,t)\delta_F(x,t)$ is singular generalized function, which is a simple layer on the surface F with specified density $\alpha = (\alpha_1, \dots, \alpha_M)$:

$$(\alpha(x,t)\delta_F(x,t), \varphi(x,t)) = \int_F \alpha_i(x,t)\varphi_i(x,t)dS(x,t), \forall\varphi(x,t) \in D_M\left(R^{N+1}\right)$$

$dS(x,t)$ is the differential of the surface in a point (x,t) and $(\nu, \nu_t) = (\nu_1, \dots, \nu_N, \nu_t)$ is a unit vector, normal to characteristic surface F.

If $F(x,t) = 0$ is an equation of wave front, then

$$(\nu, \nu_t) = (gradF, F_{,t})/\|(gradF, F_{,t})\|.$$

If the right part of expression (7) is equal to zero, then the function $\hat{u}(x,t)$ will satisfy to the Eq. (1) in a generalized sense. The natural requirement of the continuity of the solutions at transition through wave front F

$$[u_i(x,t)]_F = 0 \tag{8}$$

vanishes only two last composed right parts of Eq. (7). Hence, it is necessary that

$$\left[\sigma_i^m \nu_m - \nu_t u_{i,t}\right]_F = 0 \tag{9}$$

These conditions on the appropriate mobile wave front F_t we can write down with the account Eq. (4). By virtue of continuity of function $u(x,t)$ for $(x,t) \in F_t$, we have

$$\begin{aligned}[f(x,t)]_F &= \lim_{\varepsilon\to+0}\left(f(x+\varepsilon\nu, t+\varepsilon\nu_t) - f(x-\varepsilon\nu, t-\varepsilon\nu_t)\right)\\ &= \lim_{\varepsilon\to+0}\left(f(x+\varepsilon n, t) - f(x-\varepsilon n, t)\right) = [f(x,t)]_{F_t};\end{aligned}$$

therefore the condition (5) is equivalent to (8).
If $(x,t) \in F_t$, then $(x + cn\Delta t, t + \Delta t) \in F_{t+\Delta t}$. Therefore,

$$F(x + cn\Delta t, t + \Delta t) - F(x,t) = \left(c\left(F_{,j}, n_j\right) + F_{,t}\right)\Delta t = 0$$

From here, we have

$$c = -F_{,t}/\left(F_{,j}, n_j\right) = -\nu_t/\sqrt{\nu_i \nu_i}$$

By virtue of it, the condition (9) will be transformed to the kind (6), where c, for each front, coincides with one of c_k. The theorem has been proved.

Corollary. On the wave fronts

$$[n_l u_{i,t} + c u_{i,l}]_{F_t} = 0, \quad i = \overline{1,M}, \quad l = \overline{1,N} \tag{10}$$

The proof follows from the condition of continuity (5). The expression (10) is the condition of the continuity of tangent derivative on the wave front.

In the physical problems of solid and media, the corresponding condition (6) is a condition for conservation of an impulse at fronts. This condition connects a jump of velocity at a wave fronts with stresses jump. By this cause, such surfaces are named as *shock wave fronts*.

Definition 1. The solution of Eq. (1), $u(x,t)$, is named as classical one if it is continuous on R^{N+1}, twice differentiable almost everywhere on R^{N+1}, and has limited number of piecewise smooth wave fronts on which conditions jumps (5) and (6) are carried out.

3. Fundamental matrices

3.1 The Green's matrix of second-order system of hyperbolic equations

Let us construct fundamental solutions of Eq. (1) on $D'_M\left(R^{N+1}\right)$.

Definition 2. $U_{jk}(x,t)$ is the Green's matrix of Eq. (1) if it satisfies to equations

$$L_{ij}(\partial_x, \partial_t) U_{jk}(x,t) + \delta_{ik}\delta(x)\delta(t) = 0, \quad i,j,k = \overline{1,M} \tag{11}$$

and next conditions:

$$U_{jk}(x,t) = 0 \quad for \quad t < 0, \forall x, \tag{12}$$

$$U_{jk}(x,0) = 0 \quad for \quad x \neq 0 \tag{13}$$

Here, by definition,

$$(\delta_{ik}\delta(x,t), \varphi_i(x,t)) = \varphi_k(0,0) \quad \forall\varphi \in D'_M\left(R^{N+1}\right)$$

For construction of Green's matrix, it is comfortable to use Fourier transformation, which brings Eq. (11) to the system of linear algebraic equations of the kind

$$L_{jk}(-i\xi, -i\omega)\overline{U}_{kl}(\xi,\omega) + \delta_{jl} = 0, \quad j,k,l = \overline{1,M}$$

Here, $(\xi,\omega) = (\xi_{1,\dots,}\xi_{N,}\omega)$ is the Fourier variables appropriate to (x,t).

By permitting the system, we receive transformation of Green's matrix which by virtue of differential polynomials uniformity looks like:

$$\overline{U}_{jk}(\xi,\omega) = Q_{jk}(\xi,\omega)Q^{-1}(\xi,\omega) \tag{14}$$

where Q_{jk} are the cofactors of the element with index (k,j) of the matrix $\{L(-i\xi, -i\omega)\}$; and Q is the symbol of operator L:

$$Q(\xi,\omega) = \det\{L_{kj}(-i\xi, -i\omega)\}$$

There are the following relations of symmetry and homogeneous:

$$Q_{jk}(\xi,\omega) = Q_{jk}(-\xi,\omega) = Q_{jk}(\xi,-\omega), Q(\xi,\omega) = Q(-\xi,\omega) = Q(\xi,-\omega) \tag{15}$$

$$Q_{jk}(\lambda\xi,\lambda\omega) = \lambda^{2M-2}Q_{jk}(\xi,\omega), Q(\lambda\xi,\lambda\omega) = \lambda^{2M}Q(\xi,\omega) \tag{16}$$

By virtue of strong hyperbolicity characteristic equation,

$$Q(\xi,\omega) = 0$$

has $2M$ roots. It is a singular matrix. There is not a classic inverse Fourier transformation of it. It defines the Fourier transformation of the full class of fundamental matrices which are defined with accuracy of solutions of homogeneous system (1). Components of this matrix are not a generalized function. To calculate the inverse transformation, it is necessary to construct regularisation of this matrix in virtue of properties (12) and (13) of Green tensor. The following theorems has been proved [10]:

Theorem 3.1. *If* c_q $\left(q = \overline{1,M}\right)$ *are unitary roots of Eq. (4), then the Green's matrix of system (1) has form*

$$U_{jk}(x,t) = \sigma_N H(t) \sum_{q=1}^{M} \int_{\|e\|=1} A_{jk}(e, c_q)$$
$$\times \left\{ \left((e,x) + c_q(e)t - i0\right)^{1-N} - \left((e,x) - c_q(e)t - i0\right)^{1-N} \right\} dS(e)$$

where $\sigma_N = (2\pi i)^{-N}(N-2)!$, $A_{jk}(e,c_q) = Q_{jk}(e,c_q)/2(c_q Q_{mm}(e,c_q))$, and $H(t)$ *is Heaviside's function.*

Theorem 3.2. *If* c_q $\left(q = \overline{1,M}\right)$ *are roots of Eq. (4) with multiplicity* m_q, *then the Green's matrix of system (1) has form*

$$U_{jk}(x,t) = \sigma_N H(t) \sum_q m_q \int_{R^N} Q_{jk,\omega}^{(m_q-1)}(e, c_q) \left(Q,_{\omega}^{(m_q)}(e, c_q) \right)^{-1}$$

$$\times \left\{ ((e,x) + c_q(e)t - i0)^{1-N} - ((e,x) - c_q(e)t - i0)^{1-N} \right\} dS(e)$$

Here, the top index in brackets designate the order of derivative on ω.

So, the construction of a Green's matrix is reduced to the calculation of integrals on unit sphere. For odd N, these theorems allow to build the Green's matrix ε-approach only. For even N and for ε−approach, it is required to integrate multidimensional surface integral over unit sphere. However, in a number of cases, this procedure can be simplified.

We notice that if the original of Q^{-1} is known, i.e.

$$J(x,t) = F^{-1}\left[Q^{-1}(\xi, \omega)\right],$$

which is built in view of conditions (12), then it is easy to restore the Green's matrix

$$U_{jk}(x,t) = Q_{jk}(i\partial_x, i\partial_t) J(x,t) \tag{17}$$

In the case of invariance of Eq. (1) relative to group of orthogonal transformations, a symbol of the operator L_{ij} is a function of only two variables $\|\xi\|, \omega$ and can be presented in the form:

$$Q(\xi, \omega) = (i\omega)^{2M} q\left(\|\xi\| \omega^{-1}\right). \tag{18}$$

It essentially simplifies the construction of the original using the Green's functions of classical wave equations. For this purpose, it is necessary to spread out $Q^{-1}(\xi, \omega)$ on simple fractions. In the case of simple roots,

$$Q(\xi, \omega) = \prod_{k=1}^{M} \left(\|\xi\|^2 - \omega^2 / c_k^2 \right)$$

$$Q^{-1}(\xi, \omega) = (-i\omega)^{-2M+2} \sum_{k=1}^{M} A_k \left(\|\xi\|^2 - \omega^2 / c_k^2 \right)^{-1} \tag{19}$$

where A_k is the decomposition constant. It is easy to see that summand in round brackets under summation sign is the symbol of the classical wave operator

$$D_k = c_k^{-2} \partial_t^2 - \Delta_N.$$

Here, Δ_N is the Laplacian for which the Green's function $U_N(x,t)$ has been investigated well [11].

From Theorem 3.1 follows the support $U_N(x,t,c)$ is:

$$K_c^+ = \{(x,t) : \|x\| \leq ct, t > 0\}$$

in R^{N+1} for even N and it is sound cone

$$K_c = \{(x,t) : \|x\| = ct, t > 0\}$$

for odd N.

For example, U_3 is the simple layer on a cone [10] and it is the singular generalized function. In this case, $J(x,t)$ is convolution over t Green's function with $H(t)$:

$$J(x,t) = \sum_{k=1}^{M} A_k(H(t) *_t ... (H(t) *_t U_N(x,t,c_k))). \quad (20)$$

Here, the convolution over t undertakes $(2M - 2)$ time, which exists, by virtue of, on semi-infinite at the left of supports of functions [11]. It is easy to check up that the boundary conditions (12) and (13) are carried out as $U_N(x,t,c)$ which satisfies them. We formulate this result as:

Theorem 3.3. *If the symbol of the operator L is presented in form* (18) *and c_k are simple roots of Eq.* (4), *then $U_{jk}(x,t)$ is defined by the formula* (17), *where $J(x,t)$ looks like* (20).

If c_k have multiplicity m_k in decomposition as (20), degrees $\left(\|\xi\|^2 - \omega^2/c_k^2\right)^{-m}$ $\left(m = \overline{1, m_k}\right)$ can appear. Using the property of convolution transformation, we receive their original in kind of complete convolution over (*x*,*t*):

$$F^{-1}\left[\left(\|\xi\|^2 - \omega^2/c_k^2\right)^{-m}\right] = (U_N(x,t,c) * ..._m * U_N(x,t,c))$$

Then, the procedure of construction of a Green's matrix is similar to the described one.

We notice that as follows from (20) in a case of N = 1, 2, the convolution operation is reduced to calculate regular integrals of simple kind:

$$U_N(x,t) *_t H(t) = \int_0^t U_N(x,t-\tau)d\tau$$

$$U_N(x,t) *_t U_N(x,t) = \int_{R^N} dV(y) \int_0^t U_N(x-y,t-\tau)U_N(y,\tau)d\tau$$

But already for N = 3 and more, the construction of convolutions is non-trivial, and for their determination, its definition in a class of generalized functions should be used.

For any regular function $\hat{G} \in D'_M(R^{N+1}) : \sup_t \ \hat{G} \in (0,\infty)$, the appropriate solution of Eq. (1) looks like the convolution

$$\hat{u}_i = U_{ik} * \hat{G}_k.$$

For regular functions, it has integral representation in form of retarded potential:

$$\hat{u}_i(x,t) = H(t) \int_0^\infty d\tau \int_{R^N} U_{ik}(x-y,\tau)G_k(y,t-\tau)dV(y)$$

If Eqs. (1) are invariant, concerning the group of orthogonal transformations, then c_k do not depend on n. In physical problems, the isotropy of medium is reduced to the specified property.

3.2 The Green's tensor of elastic medium

For isotropic elastic medium constants, the matrix is equal to

$$C_{ij}^{ml} = \rho\left\{\lambda\delta_l^m\delta_i^j + \mu\left(\delta_i^m\delta_j^l + \delta_j^m\delta_i^l\right)\right\}.$$

The coefficients of Eq. (1) depend only on two sound velocities

$$c_1 = \sqrt{(\lambda + 2\mu)/\rho}, \ c_2 = \sqrt{\mu/\rho},$$

where ρ is the density of medium, and λ and μ are elastic Lame parameters. These two speeds are velocities of propagation of dilatational and shearing waves. Wave fronts for Green's tensor are two spheres expanding with these velocities.

In the case of plane deformation N = M = 2, an appropriate Green's tensor was constructed in [5, 6]. For the space deformation N = M = 3, the expression of a Green's tensor was represented in [6].

For anisotropic medium in a plane case (N = M = 2), the Green's tensor was constructed in [12, 13]. For such medium, the wave propagation velocities depend on direction n and the form of wave fronts essentially depends on coefficients of Eq. (1). Anisotropic mediums with weak and strong anisotropy of elastic properties in the case of plane deformation were considered in [12–15]. In the first case, the topological type of wave fronts is similar to extending spheres. In the second case, the complex wave fronts and *lacunas* appear [16]. *Lacunas* are the mobile unperturbed areas limited by wave fronts and extended with current of time. Such medium has sharply waveguide properties in the direction of vector of maximal speeds. The wave fronts and the components of Green's tensor for weak and strong anisotropy are presented in [15]. The calculations are carried out for crystals of aragonite, topaz and calli pentaborat.

3.3 The fundamental matrices $\hat{V}, \hat{T}, \hat{W}, \hat{U}^{(s)}, \hat{T}^{(s)}$

For solution of BVP using Green's matrix $\hat{U}$, we introduce the fundamental matrices $\hat{S}$ and $\hat{T}$ with elements given by

$$\hat{S}_{ik}^{m}(x,t) = C_{ij}^{ml}\partial_l\hat{U}_j^k, \qquad \Gamma_i^k(x,t,n) = \hat{S}_{ik}^{m}n_m, \tag{21}$$

$$\hat{T}_k^i(x,t,n) = -\Gamma_i^k(x,t,n) = -C_{ij}^{ml}n_m\partial_l\hat{U}_j^k, \tag{22}$$

$$i,j,k = \overline{1,M}, \qquad m,l = \overline{1,N}.$$

Then, the equation for $\hat{U}$ can be written as

$$\hat{S}_{ik,l}^{l} - \hat{U}_{i,tt}^{k} + \delta_i^k\delta(x)\delta(x) = 0.$$

From the invariance of the equations for $\hat{U}$ under the symmetry transformations $y = -x$, some symmetry properties of introduced matrices follows:

$$\hat{U}_i^k(x,t) = \hat{U}_i^k(-x,t), \quad \hat{U}_i^k(x,t) = \hat{U}_k^i(x,t), \quad \hat{S}_{ik}^{m}(x,t) = -\hat{S}_{ik}^{m}(-x,t), \tag{23}$$

$$\hat{T}_i^k(x,t,n) = -\hat{T}_i^k(-x,t,n) = -\hat{T}_i^k(x,t,-n). \tag{24}$$

Is easy to prove [17].

Theorem 3.4. *For fixed k and n, the vector* $\hat{T}_i^k(x,t,n)$ *is the fundamental solution of system (1) corresponding to*

$$G_i = C_{ik}^{ml} n_m \delta,_l(x)\delta(t).$$

The matrix $\hat{T}$ is called a *multipole matrix*, since it describes the fundamental solutions of system (1) generated by concentrated multipole sources (see [18]).

Primitives of the matrix. The primitive of the multipole matrix is introduced as convolution over time:

$$\hat{W}_j^k(x,t,n) = \hat{T}_j^k(x,t,n) *_t H(t),$$

which is the primitive of the corresponding matrices with respect to t:

$$\partial_t \hat{V}_i^k = \hat{U}_i^k(x,t), \quad \partial_t \hat{W}_i^k = \hat{T}_i^k(x,t,n).$$

It is easy to see that $\hat{V}_i^k$ and $\hat{W}_i^k$ are fundamental solutions to system (1) of the form

$$L_{ij}(\partial_x, \partial_t)\hat{V}_j^k + \delta_i^k \delta(x)H(t) = 0, \tag{25}$$

$$L_{ij}(\partial_x, \partial_t)\hat{W}_j^k + n_m C_{ki}^{ml} \delta,_l(x)H(t) = 0.$$

Relation (23) implies the following symmetry properties of the above matrices:

$$\hat{V}_i^k(x,t) = \hat{V}_i^k(-x,t), \quad \hat{V}_i^k(x,t) = \hat{V}_k^i(x,t),$$

$$\hat{W}_i^k(x,t,n) = -\hat{W}_i^k(-x,t,n) = -\hat{W}_i^k(x,t,-n). \tag{26}$$

The Green's matrix of the static equations for $\hat{U}_i^{k(s)}(x)$ (when the t-derivatives in (1) are zero) is defined by

$$L_{ij}(\partial_x, 0)\hat{U}_j^{k(s)}(x) + \delta_i^k \delta(x) = 0, \tag{27}$$

$$\hat{U}_i^{k(s)}(x) \to 0, \quad \|x\| \to \infty. \tag{28}$$

By analogy with (22), we define the matrix

$$\hat{T}_i^{k(s)}(x,n) = -C_{kj}^{ml} n_m \partial_l \hat{U}_j^{i(s)}.$$

Obviously, we have the symmetry relations

$$\hat{T}_i^{k(s)}(x,n) = -\hat{T}_i^{k(s)}(-x,n) = -\hat{T}_i^{k(s)}(x,-n). \tag{29}$$

Theorem 3.4 implies the following result.

Corollary. $\hat{T}_i^{k(s)}$ is a *fundamental solution of the static equations*:

$$L_{ij}(\partial_x, 0)T_j^{k(s)} - n_m C_{ki}^{ml} \delta,_l(x) = 0.$$

It is easy to see that this is an elliptic system.

The following theorem have been proved [17].

Theorem 3.5. *The following representations take place*

$$\hat{V}_i^k(x,t) = U_i^{k(s)}(x)H(t) + V_i^{k(d)}(x,t), \tag{30}$$

$$\hat{W}_i^k(x,t) = T_i^{k(s)}(x)H(t) + W_i^{k(d)}(x,t), \tag{31}$$

where $U_i^{k(s)}(x)H(t)$ and $T_i^{k(s)}(x)H(t)$ are regular functions for $x \neq 0$. As $\|x\| \to 0$,

$$U_i^{k(s)}(x) \sim \ln\|x\| A_{ik}^N(e_x), \quad T_i^{k(s)}(x) \sim \|x\|^{-1} B_{ik}^N(e_x), \quad N = 2,$$

$$U_i^{k(s)}(x) \sim \|x\|^{-N+2} A_{ik}^N(e_x), \quad T_i^{k(s)}(x) \sim \|x\|^{-N+1} B_{ik}^N(e_x), \quad N > 2. \tag{32}$$

Here, $e_x = x/\|x\|, A_{ik}^N(e),$ *and* $B_{ik}^N(e)$ *are continuous and bounded functions on the sphere* $\|e\| = 1$, *and* $V_i^{k(d)}$ *and* $W_i^{k(d)}$ *are regular functions that are continuous at* $x = 0$ *and* $t > 0$. *For any N,*

$$V_i^{k(d)}(x,t) = 0 \quad W_i^{k(d)}(x,t) = 0 \quad \textit{for } \|x\| > \max_{k=\overline{1,M}} \max_{\|e\|=1} c_k(e)t,$$

and for odd N, these relations hold for $\|x\| < \min_{k=\overline{1,M}} \min_{\|e\|=1} c_k(e)t$..

4. Statement of the initial BVP

Consider the system of strict hyperbolic equations (1). Assume that $x \in S^- \subset R^N$, where S^- is an open bounded set; $(x,t) \in D^-$, $D^- = S^- \times (0,\infty)$, $D_t^- = S^- \times (0,t)$, $t > 0$; $D = S \times (0,\infty)$, and $D_t = S \times (0,t)$.

The boundary S of S^- is a Lyapunov surface with a continuous outward normal $n(x)$ $(\|n\| = 1)$:

$$\|n(x_2) - n(x_1)\| = O(\|x_2 - x_1\|^\beta), \ \beta > 0, \ x_1 \in S, \ x_2 \in S.$$

It is assumed that G is a locally integrable (regular) vector function.

$$G \to 0 \text{ as } t \to +\infty, \quad \forall x \in S^-.$$

Furthermore, $u \in C(D^- + D)$, where u is a twice differentiable vector function almost everywhere on D^-, except for possibly the characteristic surfaces (F) in R^{N+1}, which correspond to the moving wavefronts (F_t) R^N. On them, conditions (5) and (6) are satisfied.

It is assumed that the number of wavefronts is finite and each front is almost everywhere a Lyapunov surface of dimension $N-1$.

Problem 1. Find a solution of system (1) satisfying conditions (5)–(7) if the boundary values of the following functions are given:

the initial values

$$u_i(x,0) = u_i^0(x), \quad x \in S^- + S \tag{33}$$

$$u_{i,t}(x,0) = u_i^1(x), \quad x \in S^-; \tag{34}$$

the Dirichlet conditions

$$u_i(x,t) = u_i^S(x,t), \quad x \in S, \quad t \geq 0; \tag{35}$$

and the Neumann-type conditions

$$\sigma_i^l(x,t)n_l(x) = g_i(x,t), \quad x \in S, \quad t \geq 0, \quad i = \overline{1,N}. \tag{36}$$

Problem 2. Construct resolving boundary integral equations for the solution of the following boundary value problems.

Initial-boundary value problem I. Find a solution of system (1) that satisfies boundary conditions (33)–(35) and front conditions (5)–(7).

Initial-boundary value problem II. Find a solution of system (1) that satisfies boundary conditions (33), (34), and (36) and front conditions (5)–(7).

These solutions are called *classical*.

Remark. Wavefronts arise if the initial and boundary data do not obey the compatibility conditions

$$w_i(x,0) = u_i^0(x), \quad u_i^S,_t(x,0) = u_i^1(x), \quad x \in S.$$

In physical problems, they describe shock waves, which are typical when the external actions (forces) have a shock nature and are described by discontinuous or singular functions.

5. Uniqueness of solutions of BVP

Define the functions

$$W(u) = 0,5C_{ij}^{ml}u_i,_m u_j,_l, \quad K(u) = 0,5\|u,_t\|^2,$$
$$E(u) = K(u) + W(u), \quad L(u) = K(u) - W(u),$$

which are called the *densities of internal, kinetic, and total energy* of the system, respectively, and L is *the Lagrangian*.

Theorem 5.1. *If u is a classical solution of the Dirichlet (Neumann) boundary value problem, then*

$$\int_{D_t^-} L(u(x,t))dV(x,t) = \int_{D_t^-} G_i(x,t)u_i(x,t)dV(x,t) +$$
$$+ \int_{D_t} g_i(x,t)u_i^S(x,t)dS(x,t) - \int_{S^-} \left(u_i(x,t)u_i,_t(x,t) - u_i^0(x)u_i^1(x)\right)dV(x)$$

Here and below, $dV(x) = dx_1 \dots dx_N$, $dV(x,t) = dV(x)dt$; $dS(x)$, *and,* $dS(x,t)$ *are the differentials of the area of S and D, respectively.*

Proof. Multiplying (1) by u_i and summing the result over i, after simple algebra, we obtain the expression

$$L = \left(C_{ij}^{ml}u_j,_m u_i\right),_l - (u_i u_i,_t),_t + G_i u_i.$$

This equality is integrated over D_t taking into account the front discontinuities and using the Gauss-Ostrogradsky theorem and initial conditions (33) and (34) to obtain

$$\int_{D_t^-} L(u(x,t))dV(x,t) = \int_{D_t^-} \left(C_{ij}^{ml} u_{j},_{m} u_i\right),_l - (u_i u_{i},_t),_t dV(x,t) +$$

$$+ \int_{D_t^-} G_i(x,t) u_i(x,t) dV(x,t) = \int_{D_t} \sigma_i^l n_l(x) u_i(x,t) dS(x,t) -$$

$$- \int_S u_i u_{i},_t(x,t) - u_i u_{i},_t(x,0)) dS(x,t) + \int_{D_t^-} G_i(x,t) u_i(x,t) dV(x,t) +$$

$$+ \sum_k \int_{F_k \cap D_t^-} u_i \left[\nu_l^k \sigma_i^l(x,t) - \nu_t^k u_{i},_t(x,t)\right]_{F_k} dF_k(x,t)$$

Here, ν_l^k, *and* ν_t^k are the components of the unit normal vector to the front $F_k(x,t)$ in R^{N+1}, for which we have [17]

$$\nu_t^k = -c_k / \left(\nu_j^k \nu_j^k\right)^{1/2}, \tag{37}$$

where c_k is the velocity of the front. With the notation introduced, the relation (37) and the front condition (7) yield the assertion of the theorem.

It is easy to see that the following result holds true.

Corollary. *If* $u_i(x,0) = 0, \quad u_{i},_t(x,0) = 0$, *and*

$$\lim_{t\to+\infty} u_{i},_l \to 0, \quad \lim_{t\to+\infty} u_{i},_t \to 0, \quad x \in S^-,$$

then

$$\int_{D^-} L(u(x,t))dV(x,t) = \int_{D^-} G_i(x,t) u_i(x,t) dV(x,t) + \int_D g_i(x,t) u_i^S(x,t) dS(x,t)$$

is proved in the following theorem [17]:

Theorem 5.2. *If u is a classical solution of the Dirichlet (Neumann) boundary value problem, then*

$$\int_{S^-} (E(u,t) - E(u,0)) dV(x) =$$

$$\int_{D_t^-} G_i(x,t) u_{i},_t(x,t) dV(x,t) + \int_{D_t} g_i(x,t) u_i^S,_t(x,t) dS(x,t).$$

It is easy to see that this theorem implies the uniqueness of the solutions to the initial-boundary value problems in question.

Theorem 5.3. *If a classical solution of the Dirichlet (Neumann) boundary value problem exists and satisfies the conditions*

$$\lim_{t\to+\infty} u_{i},_l \to 0, \quad \lim_{t\to+\infty} u_{i},_t \to 0, \ \forall x \in S^-,$$

then this solution is unique.

Proof. Since the problem is linear, it suffices to prove the uniqueness of the solution to the homogeneous boundary value problem. If there are two solutions u_1

and u_2, then their difference $u = u_1 - u_2$ satisfies the system of equations with $G = 0$ and the zero initial conditions, i.e.

$$u_i^m(x) = 0 \quad (m = 0, 1).$$

The vector u on the boundary S satisfies the homogeneous boundary conditions

$$u_i(x,t) = 0 \quad \text{or} \quad g_i(x,t) = 0.$$

Theorem 5.2 yields

$$\int_S E(u,t)dS(x) = \int_S (K(u,t) + W(u,t))dS(x) = 0.$$

Since the integrand is positive definite and by the conditions of the theorem, we have $u \equiv 0$. The theorem is proved.

6. Analogues of the Kirchhoff and Green's formulas

Let us assume that S is a smooth boundary with a continuous normal of a set S^-. The characteristic function $H_S^-(x)$ of a set S^- is defined for $x \in S$ as

$$H_S^-(x) = 1/2 \tag{38}$$

The Heaviside function $H(t)$ is extended to zero by setting $H(0) = 1/2$. Define the characteristic function of D^- as

$$H_D^-(x,t) = H_S^-(x)H(t) \tag{39}$$

Accordingly, for u defined on D^-, we introduce the generalized function

$$\hat{u}(x,t) = uH_D^-(x,t), \tag{40}$$

which is defined on the entire space R^{N+1}. Similarly,

$$\hat{G}_k(x,t) = G_k H_D^-(x,t). \tag{41}$$

Let $\hat{U}_i^k(x,t)$ denotes the Green's matrix, i.e. the fundamental solution of Eq. (1) that corresponds to the function $F_i = \delta_i^k \delta(x)\delta(t)$ and satisfies the conditions

$$\hat{U}_i^k(x,0) = 0, \quad \hat{U}_{i\,,t}^k(x,0) = 0, \quad x \neq 0 \tag{42}$$

For system (1), such a matrix was constructed in [10].

The primitive of Green's matrix with respect to t is defined as

$$\hat{V}_i^k(x,t) = \hat{U}_i^k(x,t) *_t H(t) \quad \Rightarrow \quad \partial_t \hat{V}_i^k = \hat{U}_i^k. \tag{43}$$

Here and below, the star denotes the complete convolution with respect to (x,t), while the variable under the star denotes the incomplete convolution with respect to x or t, respectively. The convolution exists since the supports are semibounded with respect to t. Clearly, the convolution is the solution of Eq. (1) at $F_i = \delta_i^k \delta(x)H(t)$.

Theorem 6.1. *If $u(x,t)$ is a classical solution of the Dirichlet (Neumann) boundary value problem, then the generalized solution $\hat{u}$ can be represented as the the sum of the convolutions*

$$\hat{u}_i = U_i^k * \hat{G}_k + U_i^k \underset{x}{*} u_k^1(x)H_S^-(x)+$$
$$+\partial_t U_i^k \underset{x}{*} u_k^0(x)H_S^-(x) + U_i^k * g_k(x,t)\delta_S(x)H(t)- \tag{44}$$
$$-C_{kj}^{ml}\partial_l V_i^k * u_{j},_t(x,t)n_m(x)\delta_S(x)H(t) - C_{kj}^{ml}\partial_l V_i^k \underset{x}{*} u_j^0(x)n_m(x)\delta_S(x).$$

Here, δ_S is a singular generalized function that is a single layer on S (see [2]), and $g_k(x,t)\delta_S(x)H(t)$ is a single layer on D.

Proof. Applying the operator L_{ij} to $\hat{u}(x,t)$, using the differentiation rules for generalized functions, and taking into account the equalities

$$\partial_j H_D^- = -n_j\delta_S(x)H(t), \partial_t H_D^- = \delta(t)H_S^-(x),$$

and the front conditions (5) and (6), we obtain

$$L_{kj}(\partial_x,\partial_t)\hat{u}_j(x,t) = \hat{G}_k(x,t) + u_k^1(x)H_S^-(x)\delta(t)+$$
$$+u_k^0(x)H_S^-(x)\dot{\delta}(t) + g_k(x,t)\delta_S(x)H(t) - C_{kj}^{ml}\{u_j(x,t)n_m(x)\delta_S(x)H(t)\},_l$$

Next, we use the properties of Green's matrix to construct a weak solution of Eq. (1) in the form of the convolution

$$\hat{w}_i(x,t) = U_i^k * \hat{G}_k + \hat{U}_i^k \underset{x}{*} u_k^1(x)H_S^-(x) + \partial_t\hat{U}_i^k \underset{x}{*} u_k^0(x)H_S^-(x)+$$
$$+\hat{U}_i^k * g_k(x,t)\delta_S(x)H(t) - C_{kj}^{ml}\hat{U}_i^k * (u_j(x,t)n_m(x)\delta_S(x)H(t)),_l. \tag{45}$$

The last convolution can be transformed using the relation (43) and applying the differentiation rules for convolutions and generalized functions:

$$C_{kj}^{ml}\partial_t\hat{V}_i^k * (u_j n_m(x)\delta_S(x)H(t)),_l = C_{kj}^{ml}\partial_l\hat{V}_i^k * (u_j n_m(x)\delta_S(x)H(t)),_t =$$
$$= C_{kj}^{ml}\partial_l\hat{V}_i^k * \left(u_j,_t n_m(x)\delta_S(x)H(t) + u_j^0(x)n_m(x)\delta_S(x)\delta(t)\right) =$$
$$= C_{kj}^{ml}\partial_t\hat{V}_i^k * u_j,_t n_m(x)\delta_S(x)H(t) + C_{kj}^{ml}\partial_l\hat{V}_i^k \underset{x}{*} u_j^0(x)n_m(x)\delta_S(x)$$

Let us show that $\hat{w}_i(x,t) = \hat{u}_i(x,t)$. Indeed, $\forall\varphi\in D_N\left(R^{N+1}\right)$

$$(\hat{w}_i,\varphi_i) = \left(\hat{U}_i^k * \hat{F}_k,\varphi_i\right) = \left(\hat{U}_i^k * L_{kj}(\partial_x,\partial_t)\hat{u}_j,\varphi_i\right) =$$
$$= \left(L_{kj}(\partial_x,\partial_t)\hat{U}_i^k * \hat{u}_j,\varphi_i\right) = \left(\delta_i^j\delta(x,t) * \hat{u}_j,\varphi_i\right) = (\hat{u}_i,\varphi_i).$$

Here, $\hat{F}_k$ denotes the right-hand side of (44). Since $(\hat{u}_i,\varphi_i) = 0$, if *supp* $\varphi\in D^+$, it follows that $\hat{w}_i(x,t) = 0$, $x\notin D^-$. This implies the assertion of the theorem, since the solution of the problem is unique.

Given initial and boundary values (33)–(36), the above formula recovers the solution in the domain. For this reason, it can be called an analogue of the Kirchhoff and Green formulas for solutions of hyperbolic systems (1). It gives a weak solution of the problems.

To represent this formula in integral form and use it for the construction of boundary integral equations for solutions of the initial-boundary value problems, we examine the properties of the functional matrices involved.

7. Singular boundary integral equations

Lemma 7.1 (analogue of the Gauss formula). *If S is an arbitrary closed Lyapunov surface in* R^N, *then*

$$\int_S T_k^{i(s)}(y-x,n(y))dS(y) = \delta_k^i H_S^-(x)$$

For $x \in S$, *the integral is singular and is understood in the sense of its principal value.*

Proof. Convolution Eq. (27) with $H_S^-(x)$ and using the differentiation rules for convolutions yields

$$L_{ij}(\partial_x, 0)U_j^{k(s)} * H_S^-(x) + \delta_k^i H_S^-(x) =$$

$$= -C_{ij}^{ml} U_{jk},_{l} * n_m \delta_i^k H_S^-(x) = \int_S T_k^{i(s)}(x-y,n(y))dS(y) + \delta_k^i H_S^-(x) = 0$$

Using (29), we obtain the formula in the lemma. Since $T_k^{i(s)}$ is regular for $x \notin S$, the formula holds for such x. Let us prove the validity of this formula for boundary points.

Let $x \in S$. Define $O_\varepsilon(x) = \{y \in S : \|y-x\| < \varepsilon\}, S_\varepsilon(x) = S - O_\varepsilon(x), \quad \Gamma_\varepsilon(x) = \{y : \|y-x\| = \varepsilon\}, \quad \Gamma_\varepsilon^-(x) = \Gamma_\varepsilon(x) \cap S^-, and \quad \Gamma_\varepsilon^+(x) = \Gamma_\varepsilon(x) \cap S^+$.

Similarly, we obtain

$$\int_{S_\varepsilon} T_k^{i(s)}(y-x,n(y))dS(y) + \int_{\Gamma_\varepsilon^-} T_k^{i(s)}(y-x,n(y))dS(y) = 0$$

$$\int_{S_\varepsilon} T_k^{i(s)}(y-x,n(y))dS(y) + \int_{\Gamma_\varepsilon^+} T_k^{i(s)}(y-x,n(y))dS(y) = \delta_k^i$$

Since the outward normals to $\Gamma_\varepsilon^-(x)$ and $\Gamma_\varepsilon^+(x)$ at opposite points y^- and y^+ of the sphere $\Gamma_\varepsilon(x)$ coincide, i.e. $n(y^-) = (x-y^-)/\varepsilon = (y^+-x)/\varepsilon = n(y^+)$, while $(y^+-x) = -(y^--x)$, we take into account the asymptotics of $T_k^{i(s)}$ and, according to Theorem 3.5, sum these two equalities and pass to the limit as $\varepsilon \to 0$, to obtain equality (30) for boundary points. The lemma is proved.

For $M = 1$ and $L_{1j}(\partial_x, 0) = \partial_j \partial_j = \Delta$, this formula coincides with the Gauss formula for the double-layer potential of Laplace equation (see [2]).

Consider formula (44). Formally, it can be represented in the integral form

$$\hat{u}_k(x,t) = \int_D \left(T_k^i(x-y,n(y),t-\tau)u_i(y,t) + U_k^i(x-y,t-\tau)g_i(y,\tau)\right)dD(y,\tau) +$$

$$+ \int_{S^-} \left(U_k^i,_t(x-y,t)u_i^0(y) + U_k^i(x-y,t)u_i^1(y)\right)dV(y) + U_k^i * \hat{G}_i$$

Under zero initial conditions, this formula coincides in form with the generalized Green formula for elliptic systems. However, the singularities of Green's matrix of the wave equations prevent us from using it for the construction of solutions to boundary value problems, since the integrals on the right-hand side do not exist because T_k^i has strong singularities on the fronts. However, the primitives of the matrix introduced in Section 3 can be used to construct integral representations of formula (44).

Theorem 7.1. *If u is a classical solution of the boundary value problem, then*

$$\hat{u}_k = U_k^i(x,t) * G_i(x,t) + U_k^i(x,t) * g_i(x,t)\delta_s(x)H(t) -$$

$$-\int_S T_k^{i(s)}(x-y)u_i(y,t)dS(y) - \int_S dS(y)\int_0^t W_k^{i(d)}(x-y,n(y),t-\tau)u_{i},_t(y,\tau)d\tau -$$

$$-\int_S W_k^{i(d)}(x-y,n(y),t)u_i^0(y)dS(y) + \left(U_k^i(x,t) \ *_x\ u_i^0(y)H_S^-(x)\right),_t$$

For $x \in S$, the integral is singular and is understood in the sense of its principal value.

Proof. For even N, the integral representation (42) has the form

$$\hat{u}_k = \int_S dS(y)\int_0^t \left(U_k^i(x-y,t-\tau)g_i(y,\tau) - W_k^i(x-y,n(y),t-\tau)u_{i},_\tau(y,\tau)\right)d\tau -$$

$$-\int_S W_k^i(x-y,n(y),t)u_i^0(y)dS(y) + \partial_t\int_{S^-} U_k^i(x-y,t)u_i^0(y)dS^-(y) +$$

$$+\int_{S^-} U_k^i(x-y,t)u_i^1(y)dV(y) + \int_{D^-} U_k^i(x-y,t-\tau)G_i(y,\tau)dV(y)d\tau$$

Here, all the integrals are regular for interior points and singular for boundary points.

Remark. If N is odd, then, since U is singular, the integrals involving U are still written in the form of a convolution, which is taken according to the convolution rules depending on the form of U. For the wave equation of odd dimension, such representations were constructed in [4].

It is easy to see that, for zero initial data, the last three integrals (in the convolution) vanish.

Applying Theorem 3.5, by virtue of (31), the second term can be represented as

$$\int_S dS(y)\int_0^t W_k^i(x-y,n(y),t-\tau)d_\tau u_i(y,\tau) =$$

$$= \int_S T_k^{i(s)}(x-y)\left(u_i(y,t) - u_i^0(y)\right)dS(y)$$

$$+\int_S dS(y)\int_0^t W_k^{i(d)}(x-y,n(y),t-\tau)u_{i},_\tau(y,\tau)d\tau$$

Here, the first integral is singular for $x \in S$ and exists in the sense of its principal value by Lemma 7.1, while the second integral is regular. Then for interior points, we obtain the formula of the theorem.

Let us show that the equality holds in the sense of definition (37) for boundary points as well.

Let $x^* \in S$, $x \in S^-$ and $x \to x^*$. Then, since the convolutions containing U_k^i and $W_k^{i(d)}$ are continuous, we obtain

$$\lim_{x\to x^*} u_k(x,t) = u_k(x^*,t) =$$

$$= \lim_{x\to x^*} \int_S T_k^{i(s)}(y-x)u_i(y,t)dS(y) + \int_S W_k^{i(d)}(x^* - y, n(y), t)u_i^0(y)dS(y) -$$

$$- \int_S dS(y) \int_0^t \Big(U_k^i(x^* - y, t-\tau)g_i(y,\tau) + W_k^{i(d)}(x^* - y, n(y), t-\tau)u_{i,\tau}(y,\tau)\Big)d\tau +$$

$$+ \int_{S^-} U_k^i(x^* - y, t)u_i^1(y)dV(y) + \int_{S^-} (U_k^i(x^* - y, t)u_i^0(y))_{,t}dV(y) +$$

$$+ \int_{D^-} U_k^i(x^* - y, t-\tau)G_i(y,\tau)dV(y)d\tau$$

By Lemma 7.1, the limit on the right-hand side can be transformed into

$$\int_S T_k^{i(s)}(y - x^*)(u_i(y,t) - u_i(x^*,t))dS(y) + u_i(x^*,t)\delta_k^i =$$

$$= V.P.\int_S T_k^{i(s)}(y - x^*)u_i(y,t)dS(y) - u_i(x^*,t)V.P.\int_S T_k^{i(s)}(y - x^*)dS(y) +$$

$$+ u_i(x^*,t)\delta_k^i = V.P.\int_S T_k^{i(s)}(y - x^*)u_i(y,t)dS(y) + 0,5u_i(x^*,t)\delta_k^i$$

Adding up and combining like terms, we derive the formula of the theorem for boundary points. The theorem is proved.

The formula on the boundary yields boundary integral equations for solving initial boundary value problems.

Theorem 7.2. *The classical solution of the Dirichlet (Neumann) initial-boundary value problem for $x \in S$ and $t > 0$ satisfies the singular boundary integral equations* ($k = \overline{1, M}$)

$$0,5u_k(x,t) = U_k^i(x,t) * G_i(x,t) + U_k^i(x,t) * g_i(x,t)\delta_s(x)H(t) -$$

$$- V.P.\int_S T_k^{i(s)}(x-y)u_i(y,t)dS(y) - \int_S dS(y)\int_0^t W_k^{i(d)}(x-y, n(y), t-\tau)u_{i,t}(y,\tau)d\tau -$$

$$- \int_S W_k^{i(d)}(x-y, n(y), t)u_i^0(y)dS(y) + \Big(U_k^i(x,t) \underset{x}{*} u_i^0(y)H_S^-(x)\Big)_{,t} + U_i^k \underset{x}{*} u_k^1(x)H_S^-(x).$$

From these equations, we can determine the unknown boundary functions of the corresponding initial-boundary value problem. Next, the formulas of Theorem 7.1 are used to determine the solution inside the domain.

8. Conclusions

The solvability of the obtained systems of BIEs in a particular class of functions is an independent problem in functional analysis. These equations can be numerically solved using the boundary element method. In special cases of nonstationary boundary value problems in elasticity theory ($M = N = 2, 3$), these equations were solved in [4, 6–8].

Acknowledgements

This research is financially supported by a grant from the Ministry of Science and Education of the Republic of Kazakhstan (No. AP05132272, AP05135494 and 3487/GF4).

Author details

Lyudmila A. Alexeyeva[1,2†] and Gulmira K. Zakiryanova[1,2*†]

1 Institute of Mathematics and Mathematical Modelling, Almaty, Kazakhstan

2 Institute of Mechanics and Engineering, Almaty, Kazakhstan

*Address all correspondence to: gulmzak@mail.ru

† These authors contributed equally.

References

[1] Hermander L. The Analysis of Linear Dierential Operators. Vol. 2. Heidelberg/ Moscow: Springer-Verlag; 1983/1986

[2] Vladimirov VS. Distributions in Mathematical Physics. Moscow: Nauka; 1978

[3] Alekseeva LA. Boundary integral equation method in initial-boundary value problems for solutions of the wave equation. Differentsial'nye Uravneniya. 1992;**28**:1451-1454

[4] Alekseeva LA. Method of generalized functions in nonstationary boundary value problems for solutions of the wave equation. Mat. Zh (Mathematical Journal), Almaty. 2006;**6**(1(19)):16-32

[5] Alekseeva LA. Analogue of formulas of Kirchhoff and Somigliana in plane problems of elastodynamics. Prikl. Mat. Mekh. (Appl. Math. Mech.) 1991;**55**(2): 298-308

[6] Aitaliev SM, Alekseeva LA, Dil'dabaev SA, Zhandyrbaev NB. Method of Boundary Integral Equations in Dynamic Problems for Elastic Multiply Connected Bodies. Almaty: Gylym; 1992

[7] Alekseeva LA, Zakir'yanova GK. Dynamic analogues of Somigliana's formulas for unsteady elastic dynamics with an arbitrary degree of anisotropy. Prikl. Mat. Mekh.(Appl. Math. Mech.) 1994;**58**:197-202

[8] Alekseyeva LA, Dildabayev SA, Zakir'yanova GK, Zhanbyrbayev AB. Boundary integral equations method in two and three-dimensional problems of elastodynamics. Computational Mechanics. 1996;**18**(2):147-157

[9] Petrashen GI. Foundations of Mathematical Theory of Propagation of Elastic Waves. Problems of Dynamic Theory of Propagation of Seismic Waves. Vol. 18. Leningrad; 1978

[10] Alekseeva LA, Zakir'yanova GK. The Green matrix for strictly hyperbolic systems with second order derivatives. Differential Equations. 2001;**37**(4):517-523

[11] Gel'fand IM, Shilov GE. Fourier transforms of rapidly increasing functions and questions of uniqueness of the solution of Cauchy's problem. Uspekhi Matematicheskikh Nauk. 1953;**8**(6):3-54

[12] Payton RG. Two–dimensional anisotropic elastic waves emanating from a point source. Proceedings of the Cambridge Philosophical Society. 1971; **70**:191-210

[13] Zakir'yanova GK. Boundary Integral Equations for the Nonstationary Dynamics of Anisotropic Media. Almaty: Cand. Sci. (Phys. Math.) Dissertation; 1990

[14] Alekseeva LA, Zakir'yanova GK. Non-stationary equations of anisotropic media in the space of generalized functions. Gauss formula. Izv. Akad. Nauk KazSSR. Ser. Fiz.-Mat. 1990;**5**:68-73

[15] Zakir'yanova GK. Fundamental solutions of hyperbolic system of second order and their graphical representations. New Trends in Analysis and Interdisciplinary Applications. Springer; 2017. pp. 409-415. DOI: 10.1007/978-3-319-48812-752

[16] Petrovskii IS. Lectures on Partial Dierential Equations. Moscow: Fizmatgiz; 1961

[17] Alekseeva LA, Zakir'yanova GK. Generalized solutions of initial–boundary value problems for second order hyperbolic systems. Computational Mathematics and Mathematical Physics. 2011;**1**(7):1194-1207

[18] Kecs W, Teodorescu P. Introduction to Distribution Theory with Technical Applications (Editura Tehnica, Bucharest 1975). Moscow: Mir; 1978

Chapter 5

Padé Approximation to Solve the Problems of Aerodynamics and Heat Transfer in the Boundary Layer

Igor Andrianov and Anatoly Shatrov

Abstract

In this chapter, we describe the applications of asymptotic methods to the problems of mathematical physics and mechanics, primarily, to the solution of nonlinear singular perturbed problems. We also discuss the applications of Padé approximations for the transformation of asymptotic expansions to rational or quasi-fractional functions. The applications of the method of matching of internal and external asymptotics in the problem of boundary layer of viscous gas by means of Padé approximation are considered.

Keywords: asymptotic methods, Padé approximation, boundary-value problem of mathematical physics, boundary layer

1. Introduction

An important drawback of asymptotic methods is the local character of solutions obtained [1–4]. Since the constructed series are often asymptotic, a simple increase in the number of terms does not remove this drawback. Essence of the problem consists of divergence of obtained series. There exist a lot of approaches to these problems [5, 6]. The method of analytic continuation (e.g., the Euler transform or generalized Euler transform [7–12]) requires a priori information about the singularities of the searched function in the complex domain [4, 9]. These methods are useful if a large number of terms of the series are known. In this case, it is possible to use the Domb-Sykes plot [5, 8]. But usually only a few terms of asymptotic series are known, and to get information from them, the method of Padé approximations (PAs) is useful [1, 2, 5, 13–15]. PAs yield meromorphic continuations of functions defined by power series and can be used even in cases where analytic continuations are inapplicable. If a PAs converges to the given function, then roots of the denominator tend to points of singularities. One-point PAs give possibilities to improve convergence of series [16–20]. Two-point PAs (TPPAs) allow matching asymptotics in transition zones and are widely used in mechanics and physics [1, 2, 4, 14, 21–24]. Overcoming the mentioned limitations of asymptotic methods for practically important problem is the purpose of this chapter. We consider at the beginning (Section 2) the mathematical bases of asymptotic methods and the use of Padé approximants for the summations of the asymptotic series. Section 3 discusses the method of combining of internal and external asymptotics (matching method) by means of Padé approximants. In the Section 4, the methods of solving specific problems of mathematical physics and mechanics of fluid and gas are demonstrated. Section 5 presents a discussion of the obtained.

2. Mathematical background: summation of asymptotic series

2.1 Analysis of power series

We suppose that by the result of the asymptotic study, one obtains the following series:

$$f(\varepsilon) \sim \sum_{n=0}^{\infty} C_n \varepsilon^n \text{ for } \qquad \varepsilon \to 0. \tag{1}$$

As is known, the radius of convergence ε_0 series (1) is determined by the distance to the nearest singularity of the function $f(\varepsilon)$ on the complex plane. To define ε_0, the Domb-Sykes plot may be useful [8, 10]. In many cases, one can effectively use the conformal mapping of the series, a fairly complete catalog of which is given in [9]. In particular, it sometimes turns out to be a successful Euler transformation [8, 10], based on the introduction of a new variable:

$$\tilde{\varepsilon} = \frac{\varepsilon}{1 - \varepsilon/\varepsilon_0}. \tag{2}$$

Recast the function f in terms of $\tilde{\varepsilon}, f \sim \sum_{n=0}^{\infty} d_n \tilde{\varepsilon}^n$, transfer the singularity at the point $\tilde{\varepsilon} = \infty$.

A natural generalization of Euler transformation looks as follows:

$$\tilde{\varepsilon} = \frac{\varepsilon}{(1 - \varepsilon/\varepsilon_0)^{\alpha}},$$

where α is the certain number.

2.2 Padé approximants

"The coefficients of the Taylor series in the aggregate have a lot more information about the values of features than its partial sums. It is only necessary to be able to retrieve it, and some of the ways to do this is to construct a Padé approximant" [11]. Padé approximants (PAs) allow us to transform of power series to a fractional-rational function. Let us define PAs, following Baker and Graves-Morris [25].

Suppose we are given the power series:

$$f(\varepsilon) = \sum_{i=1}^{\infty} c_i \varepsilon^i, \tag{3}$$

PAs can be written as the following expression:

$$f_{[n/m]}(\varepsilon) = \frac{a_0 + a_1\varepsilon + \ldots + a_n\varepsilon^n}{1 + b_1\varepsilon + \ldots + b_m\varepsilon^m}, \tag{4}$$

whose coefficients are determined from the condition

$$(1 + b_1\varepsilon + \dots + b_m\varepsilon^m)(c_0 + c_1\varepsilon + c_2\varepsilon^2 + \dots) = a_0 + a_1\varepsilon + \dots + a_n\varepsilon^n + O(\varepsilon^{n+m+1}) \tag{5}$$

Equating coefficients near the same powers ε, one obtains a system of linear algebraic equations. In the case where this system is solvable, one can obtain the Padé coefficients of the numerator and denominator of the PAs.

We note some properties of the PAs [5, 13, 19]. If the PAs at the chosen m and n exists, then it is unique.

1. If the PAs sequence converges to some function, the roots of its denominator tend to the poles of the function. This allows for a sufficiently large number of terms to determine the pole and then perform an analytical continuation.

2. PAs gives meromorphic continuation of a given power series.

3. PAs of the inverse function is treated as the PAs function inverse itself. This property is called duality and is more exactly formulated as follows. Let

$$q(\varepsilon) = f^{-1}(\varepsilon) \text{ and } f(0) \neq 0, \text{then } q_{[n/m]}(\varepsilon) = f^{-1}_{[n/m]}(\varepsilon) \tag{6}$$

4. Diagonal PAs are invariant under fractional linear transformations of the argument. Suppose that the function is given by their expansion (3). Consider the linear fractional transformation that preserves the origin $W = a\varepsilon/(1 + b\varepsilon)$ and the function $q(W) = f(\varepsilon)$. Then $q_{[n/n]}(W) = f_{[n/n]}(\varepsilon)$, provided that one of these approximations exist. In particular, the diagonal PAs is invariant concerning Euler transformation (2).

5. Diagonal PAs are invariant under fractional linear transformations of functions. Let us analyze a function (3). Let

$$q(\varepsilon) = \frac{a + bf(\varepsilon)}{c + df(\varepsilon)}.$$

   If $c + df(0) \neq 0$, then

$$q_{[n/n]}(\varepsilon) = \frac{a + bf_{[n/n]}(\varepsilon)}{c + df_{[n/n]}(\varepsilon)}$$

   provided that there is $f_{[n/n]}(\varepsilon)$.

6. PAs can get the upper and lower bounds for $f_{[n/n]}(\varepsilon)$. For the diagonal PAs, one has the following estimate:

$$f_{[n/n-1]}(\varepsilon) \leq f_{[n/n]}(\varepsilon) \leq f_{[n/n+1]}(\varepsilon). \tag{7}$$

   Typically, this estimate is valid for the function itself, that is, $f_{[n/n]}(\varepsilon)$ in Eq. (7) can be replaced by $f(\varepsilon)$.

7. Diagonal and close to them a sequence of PAs often possesses the property of autocorrection [17, 18]. It consists of the following. To determine the coefficients of the numerator and denominator of PAs, we have to solve

systems of linear algebraic equations. This is an ill-posed procedure, so the coefficients of PAs can be determined with large errors. However, these errors in a certain sense are of self-consistent, so the accuracy of PAs is high. This is the radical difference the PAs from the Taylor series, the calculation error of which only increases with increasing number of terms.

Autocorrection property is verified for a number of special functions. At the same time, even for elliptic functions, the so-called Froissart doublets phenomenon arises [26]. Thus, in general, having no information about the location of the poles of the PAs, but relying solely on the very PAs (computed exactly as you wish), we cannot say that you have found a good approximated function. Now consider the question: In what sense the available mathematical results on the convergence of the PAs can facilitate the solution of practical problems? Gonchar's theorem [16] states: If none of the diagonal PAs $f_{[n/n]}(\varepsilon)$ has poles in the circle of radius R, then the sequence $f_{[n/n]}(\varepsilon)$ is uniformly convergent in the circle to the original function $f(\varepsilon)$. Moreover, the absence of poles of the sequence of the $f_{[n/n]}(\varepsilon)$ in a circle of radius R confirms convergence of the Taylor series in the circle. Since the diagonal PAs is invariant under fractional linear maps $\varepsilon \to \varepsilon/(a\varepsilon + b)$, the theorem is true for any open circle containing the point of decomposition, and for any area, which is the union of these circles. A significant drawback in practice is the need to check all diagonal PAs. The fact is that if a circle of radius R has no poles only for a subsequence of the sequence of diagonal PAs, then the uniform convergence to its original holomorphic in the disk is guaranteed only with $r < r_0$, where $0.583 < r_0 < 0.584$ [27]. How can we use these results? Suppose that there are a few terms of the perturbation series and one wants to estimate its radius of convergence R. Consider the interval $[0,\varepsilon_0]$, where the truncated perturbation series and the diagonal PAs of the maximal possible order differ by no more than 5% (adopted in the engineering accuracy of the calculations). If none of the previous diagonal PAs does not have in a circle of radius ε_0 poles, then it is a high level of confidence to assert that $R \geq \varepsilon_0$.

3. Matching of limiting asymptotic expansions

3.1 Method of asymptotically equivalent functions

This method was originally proposed by Slepyan and Yakovlev for the inversion of the integral transformations. Here is a description of this method, following [26].

Suppose that the Laplace transform of a function of a real variable $f(t)$ is

$$F(s) = \int_0^\infty f(t) e^{-st} ds.$$

To obtain an approximate expression for the inverse transform, it is necessary to clarify the behavior of the transform to the vicinity of the points $s = 0$ and $s = \infty$ and to determine whether the nature and location of its singular points are on the exact boundary of the regularity or near it. Then the transform $F(s)$ is replaced by the function $F_0(s)$, approximated the exact inversion and satisfying the following conditions:

1. Functions $F_0(s)$ and $F(s)$ are asymptotically equivalent at $s \to \infty$ and $s \to 0$, that is,

$$F_0(s) \sim F(s) \text{ at } s \to 0 \text{ and } s \to \infty.$$

2. Singular points of functions $F_0(s)$ and $F(s)$, located on the exact boundary of the regularity, coincide.

The free parameters of the function $F_0(s)$ are chosen so as to satisfy the conditions of the good approximation of $F(s)$ in the sense of minimum relative error for all real values $s \geq 0$:

$$\max \left| \frac{F_0(s, \alpha_1, \alpha_2, \ldots, \alpha_k)}{F(s)} - 1 \right| \to \min \tag{8}$$

Condition (8) is achieved by variation of free parameters α_k. Often the implementation of equalities

$$\int_0^\infty F_0(s)ds = \int_0^\infty F(s)ds$$

or $F_0'(s) \sim F'(s)$ at $s \to 0$ leads to a rather precise fulfillment of the requirements (8).

Constructed in such a way function $F_0(s)$ is called asymptotically equivalent function for $F(s)$ (AEF). Let's dwell on the terminology. In the following sections, we will use the symbols of ordinal relations. We will give strict definitions of these concepts.

Let's consider the function $f(x)$. To describe the ordinal relationships with respect to another function $\varphi(x)$, enter the following definitions:

Definition 1. Let us say that $f(x)$ is a value of order $\varphi(x)$ at $x \to x_0$, that is,

$$f(x) = O(\varphi(x))$$

if $\forall \delta > 0 \exists A : |x - x_0| < \delta \Rightarrow |f(x)| \leq A|\varphi(x)|$.

Definition 2. Let us say that $f(x)$ is a value of order less than $\varphi(x)$ at $x \to x_0$, that is,

$$f(x) - o(\varphi(x))$$

if $\forall \delta > 0 \exists \varepsilon : |x - x_0| < \delta \Rightarrow |f(x)| \leq \varepsilon|\varphi(x)|$.

Here A is a finite number, and ε, δ are infinitely small.

Definition 3. Let us say that $f(x)$ is asymptotically equal to $\varphi(x)$ at $x \to x_0$, that is,

$$f(x) \sim \varphi(x) \text{ if } \frac{f(x)}{\varphi(x)} \to 1.$$

Here we use the term "asymptotically equivalent function." Other terms ("reduced method of matched asymptotic expansions" [28], "quasi-fractional approximants" (QAs) [29], and "mimic function" [30]) are also used.

3.2 Two-point Padé approximants

The analysis of numerous examples confirms "complementarity principle": if for $\varepsilon \to 0$, one can construct a physically meaningful asymptotics, there is a nontrivial asymptotics and for $\varepsilon \to \infty$. The most difficult in the asymptotic approach is the intermediate case of $\varepsilon \sim 1$. In this domain, typically numerical methods work well; however, if the task is to investigate the solution depending on the parameter ε, then it is inconvenient to use different solutions in different areas. Construction of a unified solution on the basis of limiting asymptotics is not a trivial task, and for this purpose, one can use a two-point Padé approximants (TPPAs). We give the definition following [25]. Let

$$F(\varepsilon) = \sum_{i=0}^{\infty} c_i \varepsilon^i \text{ at } \varepsilon \to 0, \tag{9}$$

$$F(\varepsilon) = \sum_{i=0}^{\infty} d_i \varepsilon^{-i} \text{ at } \varepsilon \to \infty \tag{10}$$

TPPA is a rational function of the form:

$$f_{[n/m]}(\varepsilon) = \frac{a_0 + a_1\varepsilon + \ \dots\ + a_n\varepsilon^n}{1 + b_1\varepsilon + \ \dots\ + b_m\varepsilon^m}, \tag{11}$$

k coefficients which are determined from the condition

$$(1 + b_1\varepsilon + \ \dots\ + b_m\varepsilon^m)(c_0 + c_1\varepsilon + c_2\varepsilon^2 + \ \dots) = a_0 + a_1\varepsilon + \ \dots\ + a_n\varepsilon^n + O(\varepsilon^{n+m+1}) \tag{12}$$

and the remaining coefficients from a similar condition for ε^{-1}.

4. Application of Padé approximants

4.1 Using of TPPAs in boundary-value problems

For boundary-value problems, we assume that there exist two asymptotics for limit values of the parameter. In this case, the method of matching of asymptotic expansions is usually used [4]. However, for correct application of the matching method, it is necessary to know the matching point or, at least, the domain of overlapping of asymptotics. An exact description of the transition layer $0 < \varepsilon < \infty$ exists only in the cases where solutions with different behaviors on opposite sides of the layer can be matched by a special function (e.g., the Airy function).

For the matching of nonoverlapping asymptotics, a method based on TPPAs has recently been developed. In [15, 21, 23], this method was applied for the construction of thermal profiles in a boundary layer of gas. In [2, 6], this method allowed one to examine the heat exchange in hypersonic boundary layers.

Two-point Padé approximations (TPPAs) are defined in Section 3.2 [see formulas (2)–(4)]. As an example of application of TPPAs, we consider the Airy boundary-value problem [4, 10, 31]:

$$y'' - \lambda^2 xy = g(x)y \text{ as } \lambda \to \infty \tag{13}$$

with boundary conditions

$$y(0) = 1, \;\; y(\infty) = 0 \tag{14}$$

This boundary-value problem has the form in terms of Airy function $U(s)$:

$$U'' - sU = 0, \quad U(0) = 1, \quad U(\infty) = 0 \tag{15}$$

The asymptotic solution for problems (13) and (14) has the form:

$$y(\mathrm{x}) = U(\mathrm{s})\left[1 + O\left(-\lambda^{-1}\right)\right] \text{ as } s = x\lambda^{2/3}. \tag{16}$$

The interior asymptotic ($s \to 0$) has the form of a power function:

$$U^i = 1 - as + \frac{1}{6}s^3 + O\left(s^4\right) \tag{17}$$

The exterior asymptotic has the form of an exponential function:

$$U^e = bs^{-1/4} \exp\left(-\frac{2}{3}s^{1/2}\right)\left[1 - \frac{5}{48}s^{-3/2} + O\left(s^{-3}\right)\right] \tag{18}$$

as $a \cong 0.7290,\ b \cong 0.7946$.
The transition layer is defined by the domain, where $x = O\left(\lambda^{-2/3}\right)$
Airy function approaches with TPPA:

$$U_a = \frac{1 - as + \frac{2}{3}s^{3/2} - \frac{2}{3}as^{5/2} + \frac{32}{5}as^4}{1 + \frac{32}{5}\frac{a}{b}s^{12/4}} \exp\left(-\frac{2}{3}s^{3/2}\right) \tag{19}$$

The TPPA (19) preserves three terms of the asymptotics at both ends and provides accuracy with relative error:

$$\Delta = \frac{|U - U_a|}{U} \sim 1.5\%$$

Parameters a and b are obtained from the integral equations (relations). The relations (20) and (21) can be obtained by multiplying Eq. (18) by 1, s, s^2, ... and then by integrating from 0 to ∞.

$$U'' = sU \Rightarrow \int_0^\infty U''ds = \int_0^\infty sUds \Rightarrow \int_0^\infty (U')'ds = \int_0^\infty sUds \Rightarrow U'|_o^\infty = \int_0^\infty sUds$$

This is the first integral relation.

$$\int_0^\infty sUds = a \tag{20}$$

$$sU'' = s^2U \Rightarrow \int_0^\infty sU''ds = \int_0^\infty s^2Uds \Rightarrow \left| \begin{matrix} s = t, & dt = ds \\ U''ds = dV, & V = U' \end{matrix} \right| \Rightarrow U'|_0^\infty - \int_0^\infty U'ds$$

$$= \int_0^\infty s^2Uds \Rightarrow \int_0^\infty s^2Uds = 1$$

This is the next integral relation.

$$\int_0^\infty s^2 U ds = 1 \tag{21}$$

Substituting in Eqs. (20) and (21) instead of U (4) interpolation U_a (7), calculate using quadrature integration formulas a = 0.7287 and b = 0.7922.

In the same manner, integral relations with weights U, U' can be obtained by part integration. Multiplying Eq. (18) by $U,\ U',\ U''$ ..., we get after integration from 0 to ∞,

$$UU'' = sU^2 \Rightarrow \int_0^\infty UU''ds = \int_0^\infty sU^2 ds \Rightarrow \left| \begin{matrix} U = t, & dt = dU \\ U''ds = dV, & V = U' \end{matrix} \right| \Rightarrow UU'^2\Big|_0^\infty - \int_0^\infty U'^2 ds$$

$$= \int_0^\infty sU^2 ds \Rightarrow a - \int_0^\infty U'^2 ds = \int_0^\infty sU^2 ds \Rightarrow \int_0^\infty \left(U'^2 + sU^2\right) ds = a$$

This is the first integral relation for the second method of producing it:

$$\int_0^\infty \left(U'^2 + sU^2\right) ds = a \tag{22}$$

$$U'U'' = sU'U \Rightarrow \int_0^\infty U'U''ds = \int_0^\infty sU'U ds \Rightarrow \left| \begin{matrix} U' = t, & dt = U''ds \\ U''ds = dV, & V = U' \end{matrix} \right|$$

$$\Rightarrow U'^2\Big|_0^\infty - \int_0^\infty U'U''ds = \int_0^\infty sU'U ds \Rightarrow -a^2 = 2\int_0^\infty sU'U ds$$

$$\Rightarrow \left| \begin{matrix} s = t, & dt = ds \\ U'Uds = dV, & V = \dfrac{U^2}{2} \end{matrix} \right| \Rightarrow -a^2 = 2\left[\frac{sU^2}{2}\Big|_0^\infty - \int_0^\infty \frac{U^2}{2} ds \right] \Rightarrow \int_0^\infty sU^2 ds$$

$$\Rightarrow \int_0^\infty (U^2) ds = a^2$$

And this is the next integral relation for the second method of producing it:

$$\int_0^\infty U^2 ds = a^2 \tag{23}$$

Using Eq. (19), from Eqs. (22) and (23), we calculate a = 0.7277, and b = 0.7966.

From the given example, it follows that the features of the asymptotic connection method are the ambiguity of the algorithm, the freedom to choose both the form of TPPAs, integral relations, and methods for calculating the parameters of the TPPAs. The question of choosing integral relations is, in fact, a question of controlling the asymptotic approximation using weights selected to obtain integral relations. Choosing the weight allows you to achieve acceptable accuracy in a particular area of the boundary layer: a weight equal to 1 means that the uniform influence of the entire layer is taken into account; a weight equal to 1, s, s^2, ... increases the influence of the outer region of the layer; and if the desired solution U, U', U'' is chosen as the weight, then its inhomogeneity increases the influence of the local region where the inhomogeneity is concentrated.

4.2 Quasi-fractional Padé approximants (modification of TPPA)

In the illustrated example (5), Eq. (18) TPPA represents a modified (quasi-fractional) two-point Padé approximant (10) by an exponential weight function, the choice of which is dictated by a kind of exterior asymptotics. Evidently, the TPPAs are not panacea. For example, one of the "bottlenecks" of the TPPAs method is related to the presence of logarithmic components in numerous asymptotic expansions. This problem is the most essential for the TPPAs, because, as a rule, one of the limits $\varepsilon \to 0$ *or* $\varepsilon \to \infty$ for a real mechanical problem gives expansions with logarithmic terms or other complicated functions. It is worth noting that in some cases these obstacles may be overcome by using an approximate method of TPPAs' construction by tacking as limit points not $\varepsilon = 0$ *and* $\varepsilon = \infty$, but some small and large values. On the other hand, Martin and Baker [32] proposed the so-called quasi-fractional approximants (QAs). Let us suppose that we have a perturbation approach in powers of ε for $\varepsilon \to 0$ and asymptotic expansions $F(\varepsilon)$ containing, for example, logarithm for $\varepsilon \to \infty$. By definition, QA is a ratio R with unknown coefficients a_i, b_i, containing both powers of ε and $F(\varepsilon)$. We give this modification of TPPA [2, 14, 15]. Let the series give for Eq. (5). Then the modification of TPPA is represented by the irrational function:

$$F(\varepsilon) = \frac{\sum_{k=0}^{m} a_k \varepsilon^k}{\sum_{k=0}^{n} b_k \varepsilon^k} \exp\left(-\sum_{k=0}^{l} c_k \varepsilon^k\right), \tag{24}$$

where $k+1$ coefficients c_k, $(k = 0, 1, 2, \ldots)$ are determined by means of $l+1$ integral equations for function from Eqs. (20) and (21). We notice that exponential terms [multiplier in expressions (17) and (18)] give for $\varepsilon = 0$ *and* $\varepsilon = \infty$ coincidence with TPPA (19). When considering the computational aspects of the connection method, it should first be assumed that the system of equations for determining the TPPA parameters is substantially nonlinear. To solve it, we developed a modification of the method of solving nonlinear algebraic systems [4, 23, 24].

4.3 Application of TPPAs in problems of incompressible liquid and gas mechanics

Consider the Blasius equation (45), which describes laminar boundary layers on a flat plate:

$$\begin{aligned} &\varphi''' + \varphi\varphi'' = 0; \\ &\varphi(0) = \varphi'(0) = 0; \;\; \varphi'(\infty) = 2 \end{aligned} \tag{25}$$

where $\varphi(\zeta) = \psi/\sqrt{x}$, $\psi(y)$ is the stream function, $\zeta = \frac{y}{2}\sqrt{\frac{\mathrm{Re}}{x}}$ is the automodel variable, and x and y are the Cartesian coordinates such that the axis x is directed along the flow. The interior asymptotic $(\zeta \to 0)$ has the form:

$$\varphi = a_2\zeta^2 - \frac{a_2^2}{30}\zeta^5 + O(\zeta^8) \tag{26}$$

The procedure for obtaining external asymptotics is nontrivial due to the presence of logarithmic components in the main elements. We describe in detail the mechanism for obtaining and evaluating both primary and secondary members of asymptotic. From Eq. (25) follows:

$$\frac{\varphi'''}{\varphi''} = \varphi \tag{27}$$

After integration of Eq. (27) by the coordinate ζ follows:

$$\ln\left[\varphi''(\zeta)\right]\Big|_0^{\zeta} = -\int_0^{\zeta} \varphi d\zeta \Rightarrow \ln\left[\varphi''(\zeta)\right] - \ln(2a_2) = -\int_0^{\zeta} \varphi d\zeta \Rightarrow \varphi''(\zeta)$$
$$= 2a_2 \exp\left(-\int_0^{\zeta} \varphi d\zeta\right) \tag{28}$$

After reintegration of Eq. (28) by the coordinate

$$\int_0^{\zeta} \varphi''(\zeta_1) d\zeta_1 = \int_0^{\zeta} 2a_2 \exp\left(-\int_0^{\zeta_4} \varphi d\zeta_2\right) d\zeta_1$$

follows:

$$\varphi'(\zeta_1)\Big|_0^{\zeta} = \int_0^{\zeta} 2a_2 \exp\left(-\int_0^{\zeta_4} \varphi d\zeta_2\right) d\zeta_1$$

subject to boundary conditions

$$\varphi'(\zeta_1) = \int_0^{\zeta} 2a_2 \exp\left(-\int_0^{\zeta_1} \varphi d\zeta_2\right) d\zeta_1 \Rightarrow \varphi'(\zeta_1) = \int_0^{\zeta} 2a_2 \frac{\varphi(\zeta_1)\exp\left(-\int_0^{\zeta_1} \varphi(\zeta_2) d\zeta_2\right)}{\varphi(\zeta_1)} d\zeta_1$$
$$\Rightarrow \varphi'(\zeta_1) = 2a_2 \int_0^{\zeta} \frac{1}{\varphi(\zeta_1)} d\left(\exp\left(-\int_0^{\zeta_1} \varphi(\zeta_2) d\zeta_2\right)\right)$$

Let us make a limit transition $\zeta \to \infty$ in the last equation and represent the integration interval as
$[0, \ \infty) = [0, \ \zeta] \cup [\zeta, \ \infty)$ follows:

$$\varphi'(\zeta) = 2 + 2a_2 \int_{\zeta}^{\infty} \frac{1}{\varphi(\zeta_1)} d\left(\exp\left(-\int_0^{\zeta_1} \varphi(\zeta_2) d\zeta_2\right)\right)$$

We use the mean theorem in the last equation

$$\varphi'(\zeta) = 2 + 2a_2 \frac{1}{\varphi(\zeta)} \exp\left(-\int_0^{\zeta_1} \varphi(\zeta_2) d\zeta_2\right) \tag{29}$$

In the resulting equation, the first compound is the principal member of the external asymptotics. To obtain the following members of the asymptotic, we will present the function as

$$\varphi = 2\zeta - c + z$$

where $z \to 0$, if $\zeta \to \infty$. Given the last expression of the function φ, Eq. (29) is obtained as follows:

$$\varphi'(\zeta) = 2 + \frac{2a_2}{2\zeta - c + z} \exp\left(-\int_0^\zeta (2\zeta_1 - c + z)d\zeta_1\right)$$

If $z = \varphi - 2\zeta + c$, then

$$\varphi'(\zeta) = 2 + \frac{2a_2}{2\zeta - c + z} \exp\left(-\zeta^2 + c\zeta\right) \exp\left(-\int_0^\zeta (\varphi - 2\zeta_1 + c)d\zeta_1\right)$$

In the external domain, where $\zeta \to \infty$ and $z \to 0$, let us receive an exterior asymptotic:

$$\varphi'(\zeta) = 2 + \frac{2a_2 D}{2\zeta - c} \exp\left(-\zeta^2 + c\zeta\right) + o\left(\frac{1}{\zeta^2}\right) \tag{30}$$

where $D = \exp\left(-\int_0^\infty (\varphi - 2\zeta + c)d\zeta\right)$.

To calculate parameter a_2, use the procedure of Section 4.1 [see formula (20)], and using weight equal to 1:

$$a_2 = \frac{1}{2}\int_0^\infty \varphi'_a(2 - \varphi')d\zeta \tag{31}$$

At that, in external domain, $\zeta \to \infty$

$$\varphi = 2\zeta - c, \quad (z \to 0)$$

Therefore,

$$c = \int_0^\infty (2 - \varphi')d\zeta \tag{32}$$

and

$$D = \exp\left(-\int_0^\infty (\varphi - 2\zeta + c)d\zeta\right) \tag{33}$$

Type of generalized and normalized TPPA of order (4,4):

$$\varphi'_a(\zeta) = 2\left[1 - \frac{\left(1 + \alpha_1\zeta + \alpha_2\zeta^2 + \alpha_3\zeta^3 + \alpha_4\zeta^4\right)\exp\left(-\zeta^2 + c\zeta\right)}{1 + \beta_1\zeta + \beta_2\zeta^2 + \beta_3\zeta^3 + \beta_4\zeta^4}\right] \tag{34}$$

Species of TPPA taking into account four nontrivial parameters:

$$\alpha_3,\ \beta_1,\ \beta_2,\ \beta_4$$

Therefore,

$$\varphi_a'(\zeta) = 2\left[1 - \frac{(1+\alpha_3\zeta^3)\exp\left(-\zeta^2+c\zeta\right)}{1+\beta_1\zeta+\beta_2\zeta^2+\beta_4\zeta^4}\right] \tag{35}$$

Parameter values are determined using local asymptotic and TPPA in the respective domain. Taking into account the decomposition of the exponent in the internal domain, we will write down the local equality:

$$2a_2\zeta - \frac{a_2^2}{6}\zeta^4 = 2\left[1 - \frac{(1+\alpha_3\zeta^3)\left(1+c\zeta-\zeta^2+\frac{(c\zeta-\zeta^2)^2}{2}+\dots\right)}{1+\beta_1\zeta+\beta_2\zeta^2+\beta_4\zeta^4}\right] \tag{36}$$

Taking into account Eq. (33) in the external domain, we will write down the second local equality:

$$2 - \frac{2a_2D}{2\zeta-c}\exp\left(-\zeta^2+c\zeta\right) = 2\left[1 - \frac{(1+\alpha_3\zeta^3)\exp\left(-\zeta^2+c\zeta\right)}{1+\beta_1\zeta+\beta_2\zeta^2+\beta_4\zeta^4}\right]$$
$$\Rightarrow a_2D\left(1+\beta_1\zeta+\beta_2\zeta^2+\beta_4\zeta^4\right) = (2\zeta-c)\left(1+\alpha_3\zeta^3\right) \tag{37}$$

Equalizing the coefficients in Eqs. (36) and (37) at the same degrees ζ, we get

$$\alpha_3 = a_2D,\ \beta_1 = a_2+c,\ \beta_2 = -\frac{c^2}{2}+a_2(a_2+c),\ \beta_4 = 2.$$

Therefore, the TPPA has the form:

$$\varphi_a'(\zeta) = 2\left[1 - \frac{(1+a_2D\zeta^3)\exp\left(-\zeta^2+c\zeta\right)}{1+(a_2+c)\zeta+\left(a_2^2+a_2c+\frac{c^2}{2}-1\right)\zeta^2+2\zeta^4}\right] \tag{38}$$

After systems (31)–(33) are solved, we will obtain

$$\begin{aligned} a_2 &= 0.6641,\\ c &= 1.7308,\\ D &= 0.3357 \end{aligned} \tag{39}$$

By substituting (39) in (38), we get an explicit expression for the TPPA.

4.4 Combining method of interior and exterior asymptotics for boundary layer of supersonic flow in compressed viscous gas by TPPA

We consider the boundary layer in hypersonic flow of viscous gas and solve a model problem which reduces to ordinary differential equations with appropriate boundary conditions. The TPPAs parameters are calculated and relevant questions are discussed. The equations of laminar boundary layer near a semi-infinite plate in the supersonic flow of viscous perfect gas, as it is known [2, 7], can be reduced to the form:

$$\left(\varphi''\frac{\mu}{T}\right)' + \varphi\varphi'' = 0, \tag{40}$$

$$\left(\mu\frac{T'}{T}\right)' + \sigma\varphi T' + a\sigma\frac{\mu}{T}\varphi''^2 = 0 \tag{41}$$

where

$$\varphi = \frac{\psi}{\sqrt{x}} = \varphi(\zeta),\ \ T = T(\zeta),\ \ \zeta = \frac{\eta}{2\sqrt{x}},\ \ \eta = \int_0^y \frac{dy}{T},\ \ a = \frac{1}{4}M^2(\kappa - 1)$$

M is the Mach number, σ is the Prandtl number, κ is the adiabatic index, ψ is the stream function, T is the temperature, μ is the viscosity coefficient, and x and y are the Cartesian coordinates.

The boundary conditions at the wall are

$$\varphi(0) = \varphi'(0) = 0,\ \ T(0) = T_s \tag{42}$$

At external boundary of layer is

$$\varphi'(\infty) = 2,\ \ T(\infty) = 1. \tag{43}$$

Interior asymptotic expansions are for $\mu = T^n$

$$\begin{aligned} \varphi' &= 2a_2\zeta - (n-1)a_2\frac{T_1}{T_s}\zeta^2 + O(\zeta^3), \\ T &= T_s + T_1\zeta - \left(2a\sigma a_2^2 + \frac{(n-1)}{2}\frac{T_1^2}{T_s}\right)\zeta^2 + O(\zeta^3) \end{aligned} \tag{44}$$

where two constants a_2 and T_1 remain undefined.

Exterior asymptotics for $\varsigma \to \infty$

$$\begin{aligned} \ln\varphi'' &= c^2 + c\zeta + \ln A + o(1), \\ \ln(-T') &= -\sigma\zeta^2 + \sigma c\zeta + \ln B + o(1) \end{aligned} \tag{45}$$

where three constants are unknown: c, A, and B.

We solve boundary problems (40) and (41) approximately by connecting asymptotics (44) and (45) TPPA

$$\varphi_a'(\zeta) = 2\left[1 - \frac{(1 + A\zeta^3)\exp(-\zeta^2 + c\zeta)}{1 + \alpha_1\zeta + \alpha_2\zeta^2 + \alpha_4\zeta^4}\right] \tag{46}$$

$$T_a'(\zeta) = \frac{\zeta_m - \zeta}{\beta_0 + \beta_1\zeta}\exp(\sigma(-\zeta^2 + c\zeta)) \tag{47}$$

Boundary conditions (45) and (46) are satisfied if to put

$$\varphi_a(\zeta) = \int_0^\infty \varphi_a'(\zeta)d\zeta,\ \ \ T_a(\zeta) = T_s + \int_0^\infty T_a'(\zeta)d\zeta \tag{48}$$

We complement the last equalities (50) and (51) with a normalizing condition:

$$1 = T_s + \int_0^\infty T'_a(\zeta)d\zeta \tag{49}$$

Following the procedure of the previous section, we will calculate the coefficients at ζ and ζ^2 in asymptotic expansions (44) and, equating them with the corresponding expressions from Eqs. (46) and (47), we will obtain equalities, from which values $\alpha_1, \alpha_2, \alpha_4, \beta_0, \beta_1$ are expressed through a_1, c, T_1, ζ_m:

$$\begin{aligned} &\alpha_1 = a_2 + c, \\ &\alpha_2 = -\frac{1}{2}(n-1)a_2\frac{T_1}{T_s} - 1 + \frac{c^2}{2} + a_2(a_2 + c), \quad \alpha_4 = 4, \\ &\beta_0 = \frac{\zeta_m}{T_1}, \quad \beta_1 = \sigma c\frac{\zeta_m}{T_1} - \frac{1}{T_1} + (n-1)\frac{\zeta_m}{T_1^2} \end{aligned} \tag{50}$$

Three parameters in asymptotics (44) are defined in the outer region if the following condition is met:

$$\beta_1 = -1/B \tag{51}$$

A priori at large M numbers, it is known that the temperature profile is non-monotonic and has a maximum within the layer at point ς_m at which, as can be seen from the second equation of the systems (40) and (41), the following condition is used:

$$T''(\zeta_m) = -a\sigma\varphi''^2(\zeta_m) \tag{52}$$

From the convexity condition of the temperature profile in the vicinity of the point ς_m, the following equality is used:

$$(\beta_0 + \beta_1\zeta_m)a\sigma\varphi''^2_a(\zeta_m) = \exp\left(-\sigma\left(\zeta_m^2 - \zeta_m\right)\right) \tag{53}$$

Let us add the received equations with the integrated ratios received on the basis of coincidence of TPPAs (46) and (47); in this case, three members in asymptotic decompositions (50) and (51), the initial system of Eqs. (40) and (41), with boundary conditions (42) and (43), by using the technique stated in the previous sections.

$$\begin{aligned} &a_2 = \frac{1}{2}\int_0^\infty \varphi'_a(2-\varphi')d\zeta \\ &c = \int_0^\infty (2-\varphi')d\zeta \end{aligned} \tag{54}$$

The integral relation for parameter A is obtained by multiplying Eq. (40) by

$$\exp\left(\zeta^2 - c\zeta\right)$$

and integrating from 0 to ∞ taking into account Eq. (48):

$$A = 2a_2 \frac{\mu(T_s)}{T_s} - \int_0^\infty \left(\varphi - \frac{\mu(T)}{T}(2\zeta - c) \right) \varphi'' \exp\left(\zeta^2 - c\zeta\right) d\zeta \tag{55}$$

Similarly, from Eq. (41), we get

$$B = \sigma \int_0^\infty \left(T' \left(\varphi - \frac{\mu(T)}{T}(2\zeta - c) \right) + a_2 \frac{\mu(T)}{T} \varphi''^2 \right) \exp\left(\sigma\left(\zeta^2 - c\zeta\right)\right) d\zeta \tag{56}$$

Thus, the integral relations (52) and (55)–(47) form a nonlinear system of equations for determining the following parameters:

$$T_1, \ a_2, \ c, \ A, \ B.$$

Integrals of the systems (37) and (42)–(44) solution were approximated using Simpson quadrature formulas. The behavior of magnitude B proved to be highly dependent on the behavior of the exponent at large, so the integral relation had to be replaced by the local condition (52), besides controlling the behavior of the TPPA near the maximum is more important than the weight of the exponent away from the wall. Thus, instead of the value of B, we include the value among the parameters sought, and the value of B is expressed from Eqs. (50) and (51).

5. Results

As an example of TPPA (see Section 3.2) used for matching of limiting asymptotics, consider the paper by Grasman et al. [33]. They dealt with Lyapunov exponents which characterize the dynamics of a system near its attractor. For the Van der Pol oscillator:

$$\ddot{x} + \mu\dot{x}\left(x^2 - 1\right) + x = 0 \tag{57}$$

Similar to the asymptotic approximation of amplitude and period, expressions are derived for the nonzero Lyapunov exponent λ_2 for both small and large parameter μ values:

$$\lambda_2 = -\mu - \frac{1}{16}\mu^3 + \frac{263}{18432}\mu^5 + \ldots, \mu \to 0, \tag{58}$$

$$\lambda_2 = -\frac{3 + 4\ln 2}{2(3 - 2\ln 2)}\mu + \ldots, \mu \to \infty. \tag{59}$$

The overlap of these series does not take place. The authors of [33] remark: "Such an overlap comes within reach if in the regular expansion a large number of terms is included." This is not correct, because the obtained series is asymptotic; so, with increasing of number of terms, the results will be worst. So, one needs a summation procedure. Some authors [34] proposed to use PAs, but in this case one needs hundreds of perturbation series terms. That is why we use TPPA. Using two terms from expansion (58) and one term from expansion (59), one obtains

$$L \approx -\frac{\lambda_2}{\mu} = \frac{1 + 0.14\mu^2}{1 + 0.079\mu^2} \tag{60}$$

Expression (60) has a pole at $\mu = -12.66$. Below, one can see some numerical results.

In **Table 1**, the second column is made by calculation results by formula (4), the third column is made by paper data [33]. One can see that TPPA gives good result for any value of used parameter.

μ	L (4)	L (NR)
1	1.057	1.0648
5	1.513	1.4724
10	1.685	1.6358
25	1.759	1.7398
50	1.768	1.7691

Table 1.
Comparison for L of numerical results (NR) from paper by [33] with TPPA formulate (60).

In Section 4.4, the problem was solved for several variants of the Mach number and the heating temperature: $M = 5; 10; 15$, $T_s = 3; 5; 7$ of the streamlined flat plate, with constant Prandtl number values $\sigma = 0.76$, adiabatic index $\kappa = 1.4$, and two values of dynamical viscosity index $\mu = T^n : n = 1;\ \ 0.76$. When the first equation of the systems (43) and (44) is solved, it becomes independent of the second equation and can be compared with the known Blasius solution (see Section 3), which was used as a test when compared to our method [35–40]. Thus, the value of

M	T_s	ς_m	T_1
5	3	0.426	1.340
10	3	0.744	9.127
10	5	0.637	7.929
10	7	0.531	6.676
15	3	0.806	22.00
15	5	0.756	20.82
15	7	0.712	19.75

Table 2.
TPPAs parameters for different Mach numbers M*, temperature* T_S*, and* n = 1 *values.*

M	5	10	10
T_s	3	3	5
a_2	0.17	0.74	0.56
ς_m	0.729	0.714	0.798
c	1.45	1.42	1.38
T_1	0.56	8.69	8.16

Table 3.
TPPAs parameters for different Mach numbers M*, temperature* T_S*, and* n = 0.76 *values.*

the parameters according to the exact solution is equal: $a_2 = 0.664$; $c = 1.72$. Our decision gives $a_2 = 0.6641$; $c = 1.7308$. Of course, such a good match is due to the fact that these parameters are largely determined by local internal asymptotics, more precisely, derived from the function on the wall. But also within the transition area, the deviation from the exact solution does not exceed 1÷2% (for φ' and T, respectively). Design values of parameters for determining approximations (37) and (38) for $n = 1$ are given in **Table 2**.

If $n = 0.76$, this value corresponds to the physical characteristics of the air, and the constant calculation results for the approximation formulas (49) and (50) are shown in **Table 3**.

6. Conclusion

The procedure of constructing the PA is much less labor-intensive than the construction of higher approximations of perturbation theory. PA can be applied to power series but also to the series of orthogonal polynomials. PA is locally the best rational approximation of a given power series. They are constructed directly and allow for efficient analytic continuation of the series outside its circle of convergence, and their poles in a certain sense localize the singular points (including the poles and their multiplicities) of the function at the corresponding region of convergence and on its boundary. PA is fundamentally different from rational approximations with (fully or partially) fixed poles, including the polynomial approximation, when all the poles are fixed in infinity. That is the above property of PA—effectively solving the problem of analytic continuation of power series—lies at the basis of their many successful applications in the analysis and the study of applied problems. Currently, the PA method is one of the most promising nonlinear methods of summation of power series and the localization of its singular points. Including the reason why the theory of the PA turned into a completely independent section of approximation theory, and these approximations have found a variety of applications both directly in the theory of rational approximations, and in perturbation theory.

Thus, the main advantages of PA compared with the Taylor series are as follows:

1. Typically, the rate of convergence of rational approximations greatly exceeds the rate of convergence of polynomial approximation. For example, the function e^{ε} in the circle of convergence approximated by rational polynomials $P_n(\varepsilon)/Q_n(\varepsilon)$ in 4^n times better than an algebraic polynomial of degree $2n$. More tangible, it is property for functions of limited smoothness. Thus, the function $|\varepsilon|$ on the interval $[-1,1]$ cannot be approximated by algebraic polynomials so that the order of approximation was better than $1/n$, where n is the degree of polynomial. PA gives the rate of convergence $\sim \exp\left(-\sqrt{2n}\right)$.

2. Typically, the radius of convergence of rational approximation is large compared with the power series. Thus, for the function $\arctan(x)$, Taylor polynomials converge only if $|\varepsilon| \leq 1$, and PA is everywhere in $C\backslash((-i\infty, -i] \cup [i, i\infty))$.

3. PA can establish the position of singularities of the function.

TPPA allows to overcome the locality of asymptotic expansions, using only a few terms of asymptotics. Unfortunately, the situations when both asymptotic limits have the form of power expansions are rarely encountered in practice, so we have to resort to other methods of AEFs construction, for example, the method quasirational approximation which is described in [23]. The method of combination (combining method) of asymptotics by using TPPA is alternative to the well-known matching method [6]; it is useful in local domains of transition layers where asymptotics are not uniform. This method was tested on well-known problems of mathematical physics, in particular, problems of fluid dynamics. The main advantage of the method is that it has an analytic form.

Author details

Igor Andrianov[1] and Anatoly Shatrov[2,3]*

1 Institute of General Mechanics, RWTH Aachen University, Aachen, Germany

2 Institute of Mathematics and Information Systems, Vyatka State University, Kirov, Russia

3 State Medical University, Kirov, Russia

*Address all correspondence to: shatrov@vyatsu.ru

References

[1] Andrianov IV. Application of Padé approximants in perturbation methods. Advances in Mechancis. 1991;**14**(2):3-15

[2] Barantsev RG, Pashkevich DA, Shatrov AV. Heat transfer in a boundary layer of reactive gas. In: Heat/Mass Transfer MIF-2000 [in Russian]. Minsk; 2000. pp. 185-188

[3] Frost PA, Harper EY. Extended Padé procedure for constructing global approximations from asymptotic expansions: An explication with examples. SIAM Review. 1976;**18**:62-91. DOI: 10.1137/1018003

[4] Shatrov AV. Matching of interior and exterior asymptotics in boundary-value problems of mathematical physics. Journal of Mathematical Sciences. 2018; **230**(5):804-807. DOI: 10.1007/s10958-018-3794-0

[5] Andrianov IV, Awrejcewicz J. New trends in asymptotic approaches: Summation and interpolation methods. Applied Mechanics Reviews. 2001;**54**: 69-92. DOI: 10.1115/1.3097289

[6] Brezinski C. Convergence acceleration during the 20th century. Journal of Computational and Applied Mathematics. 2000;**122**:1-21. DOI: 10.1016/S0377-0427(00)00360-5

[7] Bellman R. Perturbation Techniques in Mathematics, Physics and Engineering. London: Holt, Rinehart and Winston; 1964. 118 p

[8] Hinch EJ. Perturbation Methods. London: Cambridge University Press; 1991. 160 p. DOI: 10.1017/S0022112092220644

[9] Kublanovskaya VN. Application of analytic continuation by means of change of variables in numerical analysis. Trudy Matematicheskogo Instituta im. Steklov. 1959;**53**:145-185 (in Russian)

[10] Van Dyke M. Analysis and improvement of perturbation series. The Quarterly Journal of Mechanics and Applied Mathematics. 1974;**27**:423-450 http://dx.doi.org/10.1093/qjmam/27.4.423

[11] Van Dyke M. Computer extension of perturbation series in fluid mechanics. SIAM Journal on Applied Mathematics. 1975;**28**:720-734. DOI: 10.1007/978-1-4757-9259-1_2

[12] Van Dyke M. Perturbation Methods in Fluid Mechanics. Stanford, CA: The Parabolic Press; 1975; 271 p. DOI: 10.1017/S000192400003390X

[13] Andrianov IV, Mikhlin Yu V, Tokarzhewsky S. Two-point Padé approximants and their applications to solving mechanical problems. Journal of Theoretical and Applied Mechanics. 1997;**35**(3):577-606

[14] Barantsev RG. Definition of asymptotics and system triades. In: Asymptotic Methods in the Theory of Systems [in Russian]. Irkutsk; 1980. pp. 70-81

[15] Barantsev RG, Pashkevich DA. Matching of asymptotics in transition layers. In: Asymptotic Methods in Problems of Aerodynamics and Design of Aircrafts [in Russian]. Irkutsk; 1994. pp. 67-70

[16] Gonchar AA. Some recent results in diagonal Padé approximants. In: Approximation Theory V. Boston: Academic Press; 1986. pp. 55-70

[17] Litvinov GL. Approximate construction of rational approximations and the effect of autocorrection error. Russian Journal of Mathematical Physics. 1994;**1**:313-352. arXiv:math/0101042v1

[18] Litvinov GL. Error autocorrection in rational approximation and interval

estimation. Central European Journal of Mathematics. 2003;**1**:1-33. DOI: 10.2478/BF02475663

[19] Suetin SP. Padé approximants and efficient analytic continuation of a power series. Russian Mathematical Surveys. 2002;**57**:43-141. DOI: 10.4213/rm475

[20] Suetin SP. On interpolation properties of diagonal Padé approximants of elliptic functions. Russian Mathematical Surveys. 2004;**59**: 201-202. DOI: 10.4213/rm767

[21] Alekseeva EA, Barantsev RG, Shatrov AV. Matching of thermal asymptotics in a boundary layer. Vestnik St. Petersburg University Ser. 1. 1996;**8**:96-99. ISSN: 00240850

[22] Apresyan LA. Padé approximants. Radiophysics and Quantum Electronics. 1979;**22**:449-466

[23] Shatrov AV. Combination of asymptotics in boundary-value problems of hydrodynamics. In: Proceedings of the International Conference (OFEA-2001); June 25–29, 2001; St. Petersburg. 2001. pp. 59-60

[24] Shatrov AV. Matching of interior and exterior asymptotics in transition layers of viscous liquids and gases. In: Proceedings of the VIII All-Russian Conference on Theoretical and Applied Mechanics; Perm, August 23–29, 2001; Ural Branch RAS, Ekaterinburg. 2001. pp. 602-603 (in Russian)

[25] Baker GA Jr. and Graves-Morris P. Padé Approximants. New York: Cambridge University Press, 1996. 673 p. DOI: 10.1017/CBO9780511530074

[26] Slepyan LI, Yakovlev Yu S. Integral Transforms in Non-Stationary Problems of Mechanics. Leningrad: Sudostroyenie; 1980; 343 p. (in Russian)

[27] Vyatchin AV. On convergence of Padé approximants. Moscow University Mathematics Bulletin. 1982;**37**(4):1-4

[28] Kashin AP, Kvaratskheliya TM, Maksimov MZ, Chikovani ZE. Higher a pproximations of the reduced method of matched asymptotic expansions and its convergence. Theoretical and Mathematical Physics. 1989;**78**:278-284. DOI: 10.1007/BF01017665

[29] Chalbaud E, Martin P. Two-point quasifractional approximant in physics: method improvement and application to $J_n(x)$. Journal of Mathematical Physics. 1992;**33**:2483-2486. DOI: 10.1063/1.529618

[30] Gaunt DS, Guttman AJ. Series expansions: analysis of coefficients. In: Domb C, Green MS, editors. Phase Transitions and Critical Phenomena. Vol. 3. New York: Academic Press; 1974. pp. 181-243

[31] Kruskal MD. Asymptotology. In: Drobot S, Viebrock PA, editors. Proceedings of the Conference on Mathematical Models in Physical Sciences; April 1962; Notre Dame, Indiana. Englewood Cliffs, NJ: Prentice-Hall. 1963. pp. 17-48

[32] Martin P, Baker GA Jr. Two-point quasifractional approximant in physics. Truncation error. Journal of Mathematical Physics. 1991;**32**:313-328. DOI: 10.1063/1.529304

[33] Grasman J, Verhulst F, Shagi-Di S. The Lyapunov exponents of the Van der Pol oscillator. Mathematical Methods in the Applied Sciences. 2005;**28**: 1131-1139. DOI: 10.1002/mma.606

[34] Andersen CM, Dadfar MB, Geer JF. Perturbation analysis of the limit cycle of the van der Pol equation. SIAM Journal on Applied Mathematics. 1984; **44**:881-895

[35] Bender SM, Orszag SA. Advanced Mathematical Methods for Scientists and Engineers. New York: McGraw-Hill; 1978. DOI: 10.1007/978-1-4757-3069-2

[36] Buonomo A. The periodic solution of Van der Pol's equation. SIAM Journal

of Applied Mathematics. 1999;**59**: 156-171. DOI: 10.4236/jamp.2019.71001

[37] Miller PD. Applied Asymptotic Analysis. Providence, Rhode Island: AMS; 2006; 488 p. Available from: https://bookstore.ams.org/gsm-75/5

[38] Shanks D. Nonlinear transforms of divergent and slowly convergent sequences. Journal of Mathematical Physics. 1955;**34**:1-42

[39] Vinogradov VN, Gay EV, Rabotnov NC. Analytical Approximation of Data in Nuclear and Neutron Physics. Moscow: Energoatomizdat; 1987; 128 p. (in Russian)

[40] Weniger EI. Nonlinear sequence transformations for the acceleration of convergence and the summation of divergent series. Computer Physics Report. 1989;**10**:189-371. arXiv:math/0306302v1

Chapter 6

Alternative Representation for Binomials and Multinomies and Coefficient Calculation

José Alfredo Sánchez de León

Abstract

Polynomials play an important role in many fields of mathematics as well as in other areas such as physics and engineering. Binomials and multinomies represent a special kind of polynomials, regarded as a wide frame of study by some mathematical branches such as discrete mathematics. Under this subject a novel method was recently developed that addresses the task of performing the calculation of binomial and multinomial coefficients, by means of the setting of an arrangement of sequences of summations. The document unfolded hereby aims to be an extension of that work. Through this document, firstly it will be deemed an equation resultant from that work, targeted at binomial calculations, and will be extended to the multinomial instance. Afterwards a theoretical case of study will be presented, to expose the application of this framework. And lastly an algorithm will be raised to set it up on a computer algebra system (CAS), and some practical examples will be bestowed.

Keywords: binomials, binomial coefficient calculation, multinomial, multinomies, multinomial coefficient calculation

1. Introduction

Polynomials, alongside binomials and multinomies, are frequently found in many study cases, even outside pure mathematics affairs, like [1]. Recently in [2] a novel method was developed to calculate binomial and multinomial coefficients, and at the same time, it was shown that a different way of expansion could be set by means of arrangements of summations and sequences. For this purpose three analytic formulas were raised. Formula (3), the first one, whose output depicts sequences of incremental numbers, so that after performing those sums, a numeric value is obtained that actually represents a binomial coefficient. The second one is formula (4), whose output comprised of sequences of $\{1\}$, so that after such sums are performed, the binomial coefficient was also computed; it was exactly the same value as the first one (series of unity rather than being incremental sequences). A proof for Eq. (3) was given, but not for Eq. (4); however, it could be carried out exactly the same way. Then formula (3) was taken to extend that result to the multinomial instances, through the application of a method developed there; a proof for the multinomial case was not given. Some examples were exposed to illustrate the applications of those equations. On the other hand, some things were overlooked, i.e., the proof that were just mentioned and another set of examples that could illustrate even more the applications of the obtained result.

 IntechOpen

This document comprises the extension of the results obtained from [2]. This extension will based upon formula (4): It will be explained in more detail, oriented towards the developing of an algorithm to perform the calculations; this formula will be extended to obtain a general one to achieve the calculation of multinomial coefficients, using the procedure from [2]; a proof for this general formula will be given. The theoretical application of formula (4) will be shown in some other case study, taking over a lemma from [3] and developing an alternative proof by these means. As it was mentioned above, two algorithms based on this result are worked up and were implemented as a program in a computer algebra system (CAS).
For its best understanding, this chapter was in general written in the order just mentioned above. For the reader's convenience, and in order to get a broad understanding of this chapter, it is highly advisable to give a read at the original document [2].

2. Mathematical framework

2.1 Fundament for 2−summands

Let F be a field [4], and $F[X]$ the ring of polynomials in the indeterminate x, by the generating set of

$$F[X] = \mathscr{L}\left(\left\{\bigcup_{j\in\mathbb{N}_0} x^j\right\}\bigcup\{1\}\right) \tag{1}$$

Now, denote the set of polynomials with positive coefficients of degree ∂ at most n over F, by

$$\pi_+ := \left\{\mathcal{P}_n(F)\subseteq F[X]\subset\mathbb{R}^\infty : \mathcal{P}_n(F) = \left(\sum_{1\leq\phi\in\mathbb{Z}_+\leq s} x_{\{\phi\}}\right)^n, \mathcal{P}_j(F)\bigcap\mathcal{P}_{j+1}(F) = \varnothing\right\} \tag{2}$$

and let $\mathscr{A}$ be the set of the coefficients of those polynomials:

$$\mathscr{A} := \bigcup_{\forall k,n\in\mathbb{N}} \pi_+\backslash \prod_{j\in\{1,2\}} x_j^{(2-j)n+(2j-3)k} \tag{3}$$

Definition 2.1. A subset A of the real numbers is said to be **inductive** if it contains the number 1, and for every x in A, the number $x+1$ is also in A. Let C be the collection of all inductive subsets of $\mathbb{R}$. Then the set [5] $\mathbb{Z}_+$ of **positive integers** is defined by the equation

$$\mathbb{Z}_+ := \bigcap_{A\in C} A\subset\mathbb{R}_+. \tag{4}$$

Definition 2.2. Lexicographic ordering: $\left(\underset{\text{LEX}}{>}\right)$.

We say [6] $p\underset{\text{LEX}}{>}q$, if we have the following conditions:

$$\begin{aligned} p\underset{\text{LEX}}{>}q &\Leftrightarrow \exists i\in\mathbb{Z}_+ \ni a_i\neq b_i\wedge a_{\min\{i\}}>b_{\min\{i\}} \\ &\Rightarrow \bigcup_{n\in\mathbb{Z}_+}\{\{a_n\}\backslash\{b_n\}\}>0 \end{aligned} \tag{5}$$

for some non-zero entry, i.e., for some $n\in\mathbb{Z}_+$.

Recall the mapping from [2]

$$\tau : \mathbb{N}_0 \times \mathbb{N}_0 \rightarrow \bigcup_{j=1}^{\partial\mathcal{P}_n(F) \in \pi_+} \mathscr{A}_j \subset \mathbb{Z}_+ \ni \binom{n}{k} \mapsto \tau \tag{6}$$

by the equation

$$\binom{\hat{n}}{\hat{k}} = \left\{ \sum_{\delta_{\hat{n}}=1}^{1} {}_{\hat{n}} \cdots \sum_{\delta_{\hat{k}}=1}^{(\hat{n}-\hat{k})+1\|\delta_{\hat{k}+1}} {}_{\hat{k}} \cdots \sum_{k=1}^{(\hat{n}-\hat{k})+1\|l} {}_2 \sum_{j=1}^{(\hat{n}-\hat{k})+1\|k} {}_1 \left[\sum_{i=1}^{1} {}_0 \, i_i \right]_{j,k,..,\delta_{\hat{k}},...,\delta_{\hat{n}}} \right\} \tag{7}$$

where

$$\cdot\cdot \sum_{\delta_\phi=1}^{(\hat{n}-\hat{k})+1\|\delta_{\phi+1}} {}_\phi \cdot\cdot i = \begin{cases} \sum_{\delta_\phi=1}^{(\hat{n}-\hat{k})+1} {}_\phi & \phi = \hat{k} \\ \sum_{\delta_\phi=1}^{\delta_{\phi+1}} {}_\phi & \phi \neq \hat{k} \end{cases} \tag{8}$$

where the widehat script in the binomial terms $(\hat{n}, \hat{k})$ was placed to distinguish those from the index set of the summation sequences, and provided that δ_ϕ represents the set of a sequence of consecutive characters in the lexicographic order, by the half-open interval,

$$\delta_\phi := \{[i, j, k, \ldots z, zh, zi, \ldots, +\infty) : 1 \leq \phi < +\infty; \phi \in \mathbb{Z}_+\} \subset \mathbb{Z}_+ \tag{9}$$

Under this outline, we would have the following ordering:

$$i \underset{\text{LEX}}{<} j \underset{\text{LEX}}{<} k \cdot\cdot \underset{\text{LEX}}{<} z \underset{\text{LEX}}{<} zh \underset{\text{LEX}}{<} zi \cdot\cdot \underset{\text{LEX}}{<} z \underset{\text{LEX}}{<} zzh \underset{\text{LEX}}{<} zzi \underset{\text{LEX}}{<} \cdots \tag{10}$$

It is of great importance that we could establish a counting relation between them in order to be used in an algorithm.

Lemma 2.3. ***Existence of a choice function.***

Given a collection [5] $\mathscr{B}$ of nonempty sets (not necessarily disjoint), there exists a function

$$c : \mathscr{B} \rightarrow \bigcup_{B \in \mathscr{B}} B \tag{11}$$

such that $c(B)$ is an element of B, for each $B \in \mathscr{B}$.

A bijection between δ_ϕ and $\mathbb{Z}_+$ can now be established. Consider the collection of residue classes which have a multiplicative inverse in $\mathbb{Z}/n\mathbb{Z}$:

$$(\mathbb{Z}/n\mathbb{Z})^\times = \{\overline{a} \in \mathbb{Z}/n\mathbb{Z} : \exists \overline{c} \in \mathbb{Z}/n\mathbb{Z} \ni \overline{a} \cdot \overline{b} = \overline{1}\} \subset \mathbb{Z}/n\mathbb{Z} \tag{12}$$

By [7] we have

$$(\mathbb{Z}/mn\mathbb{Z})^\times \cong (\mathbb{Z}/m\mathbb{Z})^\times \times (\mathbb{Z}/n\mathbb{Z})^\times \tag{13}$$

when m and n are relative prime integers. However, we will set $n = 19$, to define the following applications:

Let $x \in \mathbb{Z}_+$ and J be an index set; define

$$g : (\mathbb{Z}/19\mathbb{Z})^\times \rightarrow \{i, j, k, \ldots, z\} \tag{14}$$
$$\text{by } g(x) := \text{mod } (x, 19) \rightarrow \{i, j, k, \ldots, z\}$$

By the Lemma 2.3 we can construct

$$f : \mathbb{Z}_+ \to \bigcup_{j \in J} \{z\}_j$$
$$\text{by} f(x) := \bigcup_{j \in \mathscr{L}\left(\lfloor \frac{x}{19} \rfloor\right)} \{z\}_j \tag{15}$$

Then the composition $f \circ g$ will be the searched function:

$$(f \circ g) : \mathbb{Z}_+ \to \delta_\phi$$
$$\text{by } (f \circ g)(x) = \bigcup_{j \in \mathscr{L}\left(\lfloor \frac{x}{19} \rfloor\right)} \{z\}_j \bigcup \{Im(g)\} \tag{16}$$

This could be graphically represented in **Figure 1**.

Definition 2.4. A set A is said to be infinite if it is not finite [8]. It is said to be countably infinite if there is a bijective correspondence:

$$f : A \to \mathbb{Z}_+ \tag{17}$$

In the next theorem, the collection δ_ϕ is countable:

Theorem 2.5. Let $\{En\}, n = 1, 2, 3, ...,$ be a sequence of countable sets, and put

$$S = \bigcup_{n=1}^{\infty} E_n \tag{18}$$

Then S is countable [9].

Since it was possible to establish a bijection between $\mathbb{Z}_+$ and the collection δ_ϕ, and since each summation in (1) corresponds to just one element of δ_ϕ, it follows that the summands in (1) are also countable.

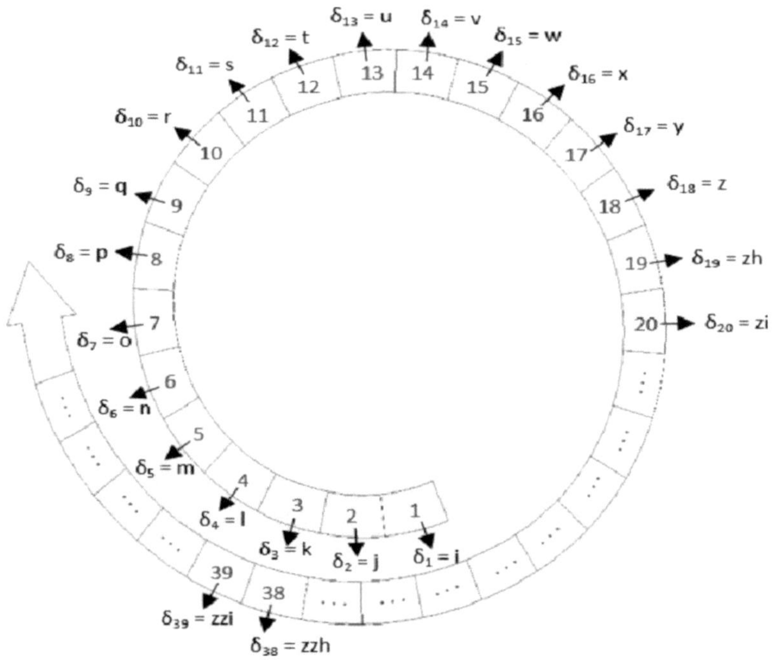

Figure 1.
Graphical representation of the mapping $(f \circ g) : \mathbb{Z}_+ \to \delta_\phi$.

Corollary 2.6. The set $\mathbb{Z}_+ \times \mathbb{Z}_+$ is countably infinite [8].

Finally, in this corollary, it follows that the tuples $\mathbb{Z}_+ \times \{\delta_\phi \subset \mathbb{Z}_+\}$ are countably infinite.

2.2 Extension to $n-$summands

Now that the countability of the collection δ_ϕ was explained and a bijective function settled down to perform it, we will proceed to extend (1) to $n-$ summands. Note that in [2] this formula remained unaltered in this sense, so that it was not extended; that is why we will do it here.

For the foremost part, another index set on a half-open interval is introduced, defined in similar way as δ_ϕ:

$$\delta_\phi^* := \{[\hat{n}, i^*, j^*, k^*, \ldots z^*, zh^*, zi^*, \ldots, +\infty) : 0 \leq \phi < +\infty; \phi \in \mathbb{Z}_+\} \subset \mathbb{Z}_+ \tag{19}$$

The need of another index set comes from the fact that in the extension to the $n-$ summands approach, a second lawyer of summands sequences and a new sequence of multipliers will arise; this way index scrambling between the two lawyers will be avoided. Then in a similar fashion, the modified function will apply:

$$(f' \circ g') : \mathbb{Z}_+ \to \delta_\phi^* \tag{20}$$

where f' and g' are defined the same way as in the former, regarding $*$ as a superscript on the alphabetic letters.

Theorem 2.7. Principle of recursive definition. Let A be a set; let a_0 be an element of A. Suppose ρ is a function that assigns to each function f mapping a nonempty section of positive integers into A, an element of A. Then there exists a unique function

$$h : \mathbb{Z}_+ \to A \tag{21}$$

such that $h(1) = a_0$,

$$h(i) = \rho(h|\{1, \ldots, i-1\}) \; \forall i > 1. \tag{22}$$

Now the same method developed in [2] will be performed, following the binomial theorem [10–12] and the recursive principle: Let $\varphi, \gamma \in \mathcal{P}_n(F) \backslash \mathscr{A}$ be two collection of summands; set the following:

$$\varphi_{\{s-f\}} = \sum_{\substack{\delta_f^* = 0|_{\lambda=0} \leq \phi_\lambda \leq \delta_{f-1}^*|_{\lambda=1} \\ 0 \leq \lambda \in \mathbb{N}_0 \leq 1 \forall \lambda}} \binom{\phi|_{\lambda=1}}{\phi|_{\lambda=0}} \gamma_f^{\phi|_{\lambda=1} - \phi|_{\lambda=0}} \varphi_{\{s-f\}}^{\phi|_{\lambda=0}} \tag{23}$$

Subindices $\{f\}, \{s-f\} \in \mathbb{Z}_+$ represent a consecutive number of a summand and the amount of remaining summands after the binomial theorem expansion, respectively. Continue recursively performing the expansion:

$$\sum_{\delta_1^*=0}^{\delta_0^*} \left\{ \sum_{\delta_{\delta_0^*}=1}^{1} {}_{\delta_0^*} \cdots \sum_{\delta_{\delta_1^*}=1}^{(\delta_0^*-\delta_1^*)+1\|\delta_{\delta_1^*+1}} {}_{\delta_1^*} \cdots \sum_{k=1}^{(\delta_0^*-\delta_1^*)+1\|l} {}_2 \sum_{j=1}^{(\delta_0^*-\delta_1^*)+1\|k} {}_1 \left[\sum_{i=1}^{1} {}_0 \, i_i \right]_{j,k,..,\delta_{\delta_1^*}, \ldots, \delta_{\delta_0^*}} \right\} \gamma_1^{\delta_0^*-\delta_1^*} \varphi_{s-1}^{\delta_1^*} \tag{24}$$

$$= \sum_{\delta_1^*=0}^{\delta_0^*}\left(\left\{\sum_{\delta_{\delta_0^*}=1}^{1}{}_{\delta_0^*}\cdots\sum_{\delta_{\delta_1^*}=1}^{(\delta_0^*-\delta_1^*)+1\|\delta_{\delta_1^*+1}}{}_{\delta_1^*}\cdots\sum_{k=1}^{(\delta_0^*-\delta_1^*)+1\|l}{}_2\sum_{j=1}^{(\delta_0^*-\delta_1^*)+1\|k}{}_1\left[\sum_{i=1}^{1}{}_0 i_i\right]_{j,k,..,\delta_{\delta_1^*},\dots,\delta_{\delta_0^*}}\right\}\gamma_1^{\delta_0^*-\delta_1^*}\bullet\right.$$

$$\sum_{\delta_2^*=0}^{\delta_1^*}\left\{\left(\sum_{\delta_{\delta_1^*}=1}^{1}{}_{\delta_1^*}\cdots\sum_{\delta_{\delta_2^*}=1}^{(\delta_1^*-\delta_2^*)+1\|\delta_{\delta_2^*+1}}{}_{\delta_2^*}\cdots\sum_{k=1}^{(\delta_1^*-\delta_2^*)+1\|l}{}_2\sum_{j=1}^{(\delta_1^*-\delta_2^*)+1\|k}{}_1\left[\sum_{i=1}^{1}{}_0 i_i\right]_{j,k,..,\delta_{\delta_2^*},\dots,\delta_{\delta_1^*}}\right\}\gamma_2^{\delta_1^*-\delta_2^*}\bullet\right.$$

$$\sum_{\delta_3^*=0}^{\delta_2^*}\left(\left\{\sum_{\delta_{\delta_2^*}=1}^{1}{}_{\delta_2^*}\cdots\sum_{\delta_{\delta_3^*}=1}^{(\delta_2^*-\delta_3^*)+1\|\delta_{\delta_3^*+1}}{}_{\delta_3^*}\cdots\sum_{k=1}^{(\delta_2^*-\delta_3^*)+1\|l}{}_2\right.\right.$$

$$\left.\sum_{j=1}^{(\delta_2^*-\delta_3^*)+1\|k}{}_1\left[\sum_{i=1}^{1}{}_0 i_i\right]_{j,k,..,\delta_{\delta_3^*},\dots,\delta_{\delta_2^*}}\right\}\gamma_3^{\delta_2^*-\delta_3^*}\varphi_{s-3}^{\delta_3^*}\bullet\ \cdots \tag{25}$$

… ••• …••• … ••• …••• … ••• …

$$\cdots\bullet\sum_{\delta_s^*=0}^{\delta_{s-1}^*}\left(\left\{\sum_{\delta_{\delta_{s-1}^*}=1}^{1}{}_{\delta_{s-1}^*}\cdots\sum_{\delta_{\delta_s^*}=1}^{(\delta_{s-1}^*-\delta_s^*)+1\|\delta_{\delta_s^*+1}}{}_{\delta_s^*}\cdots\sum_{k=1}^{(\delta_{s-1}^*-\delta_s^*)+1\|l}{}_2\right.\right.$$

$$\left.\sum_{j=1}^{(\delta_{s-1}^*-\delta_s^*)+1\|k}{}_1\left[\sum_{i=1}^{1}{}_0 i_i\right]_{j,k,..,\delta_{\delta_s^*},\dots,\delta_{\delta_{s-1}^*}}\right\}\gamma_{s-1}^{\delta_{s-1}^*-\delta_s^*}\varphi_s^{\delta_{s-1}^*}\Big)\cdots\Big)\cdots\Big)\cdot\cdots\Big)$$

Then the expansion above follows a pattern that can be coded into a general formula; replace the variables γ and φ according to its definition; it would end with the function

$$\{\pi_+\subset F[X]\}\backslash\mathscr{A}:\longrightarrow\pi_+\subset F[X]$$

$$\bigcup_{j=1}^{s}\{x_j\}\cup\{n\}\mapsto\left\{h\left(\bigcup_{j=1}^{s}\{x_j\}\cup\{n\}\right)=\left(\sum_{1\le\phi\in\mathbb{Z}_+\le s}x_{\{\phi\}}\right)^n\right\}$$

$$\left(\sum_{1\le\phi\in\mathbb{Z}_+\le s}x_{\{\phi\}}\right)^n=\prod_{f=1}^{s-1}\left[\sum_{\delta_f^*=0}^{\delta_{f-1}^*}\left(\left\{\sum_{\delta_{\delta_{f-1}^*}=1}^{1}{}_{\delta_{f-1}^*}\cdots\sum_{\delta_{\delta_f^*}=1}^{\left(\delta_{f-1}^*-\delta_f^*\right)+1\|\delta_{\delta_f^*+1}}{}_{\delta_f^*}\cdots\sum_{k=1}^{\left(\delta_{f-1}^*-\delta_f^*\right)+1\|l}{}_2\right.\right.\right. \tag{26}$$

$$\left.\left.\left.\sum_{j=1}^{\left(\delta_{f-1}^*-\delta_f^*\right)+1\|k}{}_1\left[\sum_{i=1}^{1}{}_0 i_i\right]_{j,k,..,\delta_{\delta_f^*},\dots,\delta_{\delta_{f-1}^*}}\right\}\right)x_f^{\delta_{f-1}^*-\delta_f^*}\right]\bullet\left(x_s^{\delta_{s-1}^*}\right)$$

where

$$\cdot\cdot\sum_{\delta_\phi=1}^{\left(\delta_{f-1}^*-\delta_f^*\right)+1\|\delta_{\phi+1}}{}_\phi\cdot\cdot k=\begin{cases}\displaystyle\sum_{\delta_{\delta_f^*}=1}^{\left(\delta_{f-1}^*-\delta_f^*\right)+1}{}_\phi & \phi=\delta_f^*\\ \displaystyle\sum_{\delta_{\delta_f^*}=1}^{\delta_{\phi+1}}{}_\phi & \phi\ne\delta_f^*\end{cases} \tag{27}$$

This general formula actually performs the multinomial expansion along with the calculation of the coefficients of individual terms, for an expression of $n-$summands; its proof is given on Appendix A.

Theorem 2.8. A finite product of countable sets is countable [5].

Since (2) represents finite products of countable sets (as it was exposed previously), it follows from theorem 2.8 that the sequence of multipliers in (2) is also countable.

3. Applications of the obtained results

Direct use of formula (1) is performed on Appendix C in [2], alongside the use of another general formula for multinomial expansion, similar to (2), just obtained in the last section of this document. What is exposed here are applications not covered in [2]; these are regarded as theoretical and numerical applications and are given next.

3.1 Theoretical application

A theoretical case application is exerted; the last results are performed to give an alternative proof for Lemma from [3].

Lemma 3.1. If p is a prime not dividing an integer m, then for all $n \geq 1$, the binomial coefficient $\binom{p^n m}{p^n}$ is not divisible by p.

Proof: Formula (1) will be deployed for this purpose. Since p is prime, each summand in it arises from

$$\cdot\cdot \sum_{\delta_\phi=1}^{(p^n m - p^n)+1 \| \delta_{\phi+1}} {}_{\phi} \cdot\cdot i = \begin{cases} \displaystyle\sum_{\delta_\phi=1}^{p^n(m-1)+1} {}_{\phi} & \phi = p^n \\ \displaystyle\sum_{\delta_\phi=1}^{\delta_{\phi+1}} {}_{\phi} & \phi \neq p^n \end{cases} \tag{28}$$

(since p cannot be factored any further). Now, suppose for the sake of contradiction that there exists a prime p that divides the binomial coefficient the way it is proposed:

$$\binom{p^n m}{p^n} = \sum_{\delta_{p^n}=1}^{p^n(m-1)+1} {}_{p^n} \cdots \sum_{\substack{1 \leq j \leq k \leq \cdots \leq \delta_\phi \leq (p^n m - p^n)+1 \\ \delta_{p^n m} = 1 \forall \delta_\phi}} \left[\sum_{i=1}^{1} {}_{0} i_i \right] =: \psi(p^n, m) = p \cdot x_{int} \tag{29}$$

for some integer $x_{int} \in \mathbb{Z}_+$, where $\psi(p^n, m)$ is defined to be the summation sequence function on the left side of (3).

$$\Rightarrow \sum_{\delta_{p^n}=1}^{p^n(m-1)+1} {}_{p^n} \cdots \sum_{l=1}^{m} {}_3 \sum_{k=1}^{l} {}_2 \sum_{j=1}^{k} {}_1 \left[\frac{1}{p} \right] = x_{int} \tag{30}$$

Since formula (1) represents addition of sequences of $\{1\}$, expanding the above and gathering out the factors, the following would be obtained:

$$\left(\frac{1}{p}\right)(1) + \left(\frac{1}{p}\right)\left(p^n + \sum_{i=0}^{p^n}(p^n - i) + \cdots + k_{p^n(m-1)+1}\right) = x_{int} \tag{31}$$

But above there are no summands on the left side that yields an integer on the right side and at the same time fulfills:

$$1 + p^n + \sum_{i=0}^{p^n}(p^n - i) + \cdots + k_{p^n(m-1)+1} = \psi(p^n, m) \tag{32}$$

For if it were,

$$\Rightarrow \exists z \in \mathbb{Z}_+ \ni \psi(p^n, m) = z \ \forall m, n, p \in \mathbb{Z}_+ : p \nmid m \tag{33}$$

so that, combining (4) and (5), it would be

$$\begin{aligned} &\frac{1}{p} + \left(\frac{1}{p}\right)(z-1) = x_{int} \\ &\Rightarrow \frac{\cancel{1}}{\cancel{p}} + \frac{z}{p} - \frac{\cancel{1}}{\cancel{p}} = x_{int} \\ &\Rightarrow \frac{z}{p} = x_{int} \quad \Rightarrow\Leftarrow \end{aligned} \tag{34}$$

which shows that indeed x_{int} is all about a rational number. This contradicts the hypothesis, so the binomial coefficient is not divisible by p.

□

There is also problem 18.41 in [13] where the author introduces a case of study that takes place when x and y are members of a commutative ring of characteristic p:

(Freshman exponentiation). Let p be prime. Show that in the ring $\mathbb{Z}_p$, we have

$$(a+b)^p = a^p + b^p \tag{35}$$

for all $a, b \in \mathbb{Z}_p$ [*hint:* observe that the usual binomial expansion for $(a+b)^n$ is valid in a *commutative* ring]. This one actually can alternatively be proven in a very similar way as the above lemma, so the proof is left to the reader. Some other study cases may come up that can be addressed with this result, where binomial coefficient calculation is part of their proof.

3.2 Numerical application

Two algorithms were written with the use of formulas (1) and (2); those were also implemented on two script programs written on the computer algebra system Maxima [14] and open-source software written in LISP [15] and based on a 1982 version of Macsyma [16]; of course there are many other CAS in which those can be implemented, i.e., here [17] is a great deal of one of them. The aim is to perform the expansion of a binomial by this result and perform the calculations of their individual coefficients. The first algorithm describes the calculation of binomial coefficients by formula (1), while the second is about the binomial and multinomial expansion based on formula (2); it also calculates the coefficients of the individual terms based on algorithm 1.

They are the exposed in the next two frames.

Algorithm 1: coef(n,k,o)

```
// Implements binomial coefficient calculation by
   sequences of summations,
// using formula (1).
```

input : $n = \partial\,(F\,[X])\,,\ k, o \in \{0\} \bigcup \mathbb{Z}$

output: $\binom{n}{k} \ni \sum_{k=0}^{n} \binom{n}{k} \in \mathscr{A}$

decide: Whether $o = \{0\}$. If so, then compute numerically the coefficients; otherwise perform and expansion of summations.

begin

Let $x \in \mathbb{Z}_+$;

Function *CHR(x)*

return $\left\langle CHR(x) = \bigcup_{j \in \mathscr{L}(\lfloor \frac{x}{19} \rfloor)} \{z\}_j \bigcup \left\{ (\mathbb{Z}/19\mathbb{Z})^{\times} \longrightarrow \{i, j, k, ..., z\} \right\} \right\rangle$

if $k < 0$ **or** $n < 0$ **then**

return $\left\langle \binom{k}{n} \leftarrow 0 \right\rangle$

if $k = 0$ **then**

return $\left\langle \binom{k}{n} \leftarrow \sum_{i=0}^{1} [i] \right\rangle$

$cf_+ := +[1]$

for $xi \leftarrow 1$ **to** k **do**

if $xi = k$ **then**

```
// Sequence stops.
```

$$cf_+ \leftarrow \left(\sum_{CHR(k)}^{n-k+1} \cdots \sum_{\substack{1 \le j \le k \le \cdots \le \delta_{xi} \le (n-k)+1 \\ \delta_n = 1\ \forall \delta_{xi}}} \bigcup \{cf_+\} \right)$$

else

```
// Sequence continues.
```

$$cf_+ \leftarrow \left(\cdots \sum_{CHR(k)}^{CHR(k+1)} \cdots \sum_{\substack{1 \le j \le k \le \cdots \le \delta_{xi} \le (n-k)+1 \\ \delta_n = 1\ \forall \delta_{xi}}} \bigcup \{cf_+\} \right)$$

```
// Depending upon value of o, simplify or not;
// o = 0 simplify,otherwise, don't.
```

set : simplification_flag $\leftarrow$ YES/NO

return $\left\langle \left\{ \binom{k}{n} \in \mathscr{A} \right\} \leftarrow cf_+ \right\rangle$

Algorithm 2: multinom(s,n,o)

```
// Implements full expansion of multinomies and
// calculation of multinomial coefficients.
```

input : $n = \partial\,(\pi_+ \subset F\,[X]) \ \wedge \ s -$ summands

output: $\left(\sum_{1 \le \phi \in \mathbb{Z}_+ \le s} x_{\{\phi\}} \right)^n \subset \pi_+$

decide: Whether $\partial = 0$. If so, then compute numerically the coefficients with *coef(n,k,o)*; otherwise perform an expansion of summations.

begin

Let $\phi \in \mathbb{Z}_+$;

Function *_CHR(ϕ)*

if $\phi = 0$ **then**

return $\langle$ _CHR(ϕ)=$\{n\}$ $\rangle$

else

return $\langle$ _CHR(ϕ)=

$\{n\} \quad \bigcup_{j \in \mathscr{L}(\lfloor \frac{\phi}{19} \rfloor)} \{z\}_j \bigcup \left\{ (\mathbb{Z}/19\mathbb{Z})^{\times} \longrightarrow \bigcup \{i^*, j^*, k^*, ..., z^*\} \right\} \rangle$

Let $x \in F$;

$msum := x_{\{s\}}^{indx}$

$indx \leftarrow$ *_CHR(xi)*

$indx_1 \leftarrow$ *_CHR(s − xi)*

$indx_2 \leftarrow$ *_CHR(s − (xi + 1))*

for $xi \leftarrow 1$ **to** $s - 1$ **do**

$$msum \leftarrow \sum_{indx_1 = 0}^{indx_2} \text{coef}\binom{indx_2}{indx_1} \cdot x_{\{s\}}^{indx_2} \cdot msum$$

return $\left\langle \left\{ \left(\sum_{1 \le \phi \in \mathbb{Z}_+ \le s} x_{\{\phi\}} \right)^n \in \pi_+ \right\} \leftarrow msum \right\rangle$

```
coef(n,k,o):=
/*Usage: coef(n,k,o); */
/*n: Grade of the polynomial or number that it is elevated to. */
/*k: k-th binomial term. */
/*o: Simplifying option, o = 0 yields numeric value of the calculated coefficient, */
/* with any other value, it expands the summations sequences. */
block([x,s],
        simp:true,
        chr_(x):= concat(smake(floor(x/19),"z"),ascii(mod(x,19)+104)),
        x_:"k",
        s:[ 1],
        if o = 0
            then option:true
            else option:false,
        if not integerp(n) or not integerp(k)
            then return("'n' and 'k' must be integers ..."),
        if k < 0 or n < 0
            then return(0),
        if k = 0
            then (simp:option, return
        (eval_string(sconcat("sum([ 1],i,1,",1,")")))),
        for idx:1 thru k do
        (
            if idx = k
            then
            (
                    s:sconcat("sum"," (",s, ",",chr_(idx),",",1,",",n-k+1,")")
            )
            else
            (
                    s:sconcat("sum",",(",s,",",chr_(idx),",",1,",",chr_(idx+1),")")
            )
        ),
        simp:option,
        eval_string(s)
      );
```

Figure 2.
Program: coef.

```
multinom(s,n,o):=
/*Usage: multinom(s,n,o); */
/*s: Amount of summands that has the polynomial, i.e., s=2 is about a binomial. */
/*n: Grade of the polynomial or number that it is elevated to. */
/*o: Simplifying option, o = 0 yields numeric value of the calculated coefficient, */
/* with any other value, it expands the sequences of the individual terms. */
block([ multsum],
        simp:true,
        if o = 0
            then opt:0
            else opt:1,
        _chr_(x):=
        block([],
                if x = 0
                    then return(n)
                    else return(concat("_",smake(floor(x/18),"z"),ascii(mod(x,18)+104))
                         )
        ),
        multsum:sconcat(x[ s],"^",indx_3),
        expr:"x[ 1]",
        for sx:1 thru s-1 do
        (
                expr:concat("(",expr,"+",x,"[ ",sx+1,"] ",")"),
                indx:_chr_(s-sx),
                indx_1:_chr_(s-(sx+1)),
                indx_3:_chr_(s-1),
                multsum:sconcat("sum","(","cooe__","(",indx_1,",",indx,",",opt,")","*",
                x[ s-sx],"^","(",indx_1,"-",indx,")","*",
                multsum,",",indx,",",0,",",indx_1,")")
        ),
        print(eval_string(expr)^n,"="),
        CMD_:eval_string(multsum),
        return(expand(ev(CMD_,cooe__=coef)))
    );
```

Figure 3.
Program: multinom.

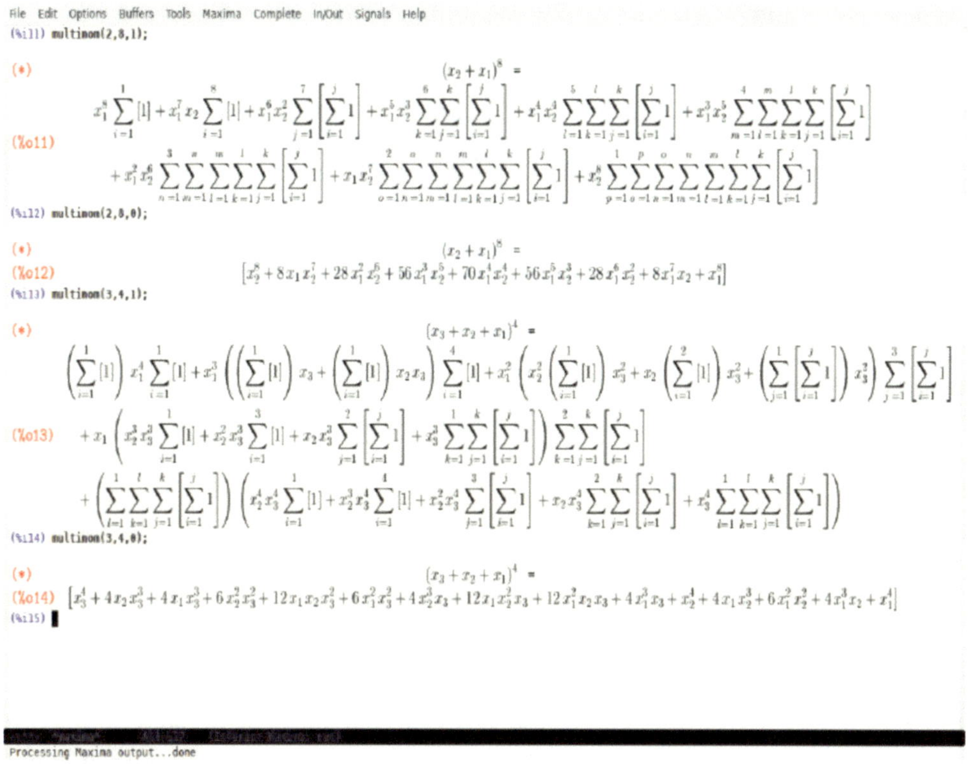

Figure 4.
Output of multinom.mc for a binomial and a multinomial example, respectively (in electronic PDF file, this figure can be zoomed fairly enough for a better view).

The implemented algorithms (1, 2) were presented above in the code displayed in **Figures 2** and **3**, for the first and the second one, respectively. To use the code, open a Maxima instance; then for the foremost part, in order to avoid LISP errors, issue the following command:

```
:lisp (setf(get '%sum 'operators) nil)
```

Next save the code above as the files *coef.mc* and *multinom.mc*, respectively, and place them in the subdirectory user of the Maxima installation (< maxima_userdir can be issued; > in the command line to display it). Following, issue the commands:

```
batch("coef");
batch("multinom");
```

in order to be both loaded; or alternatively just copy the text from **Figures 2** and **3**, and paste it in the Maxima interface. Then use them according to each program syntax. The program *multinom.mc* outputs are shown in **Figure 4**.

A pattern of outputs (binomial coefficient values) from *coef.mc* was included; this is displayed in Appendix B.

4. Conclusions

With this result an alternative way to represent binomials and multinomies alongside their respective coefficient calculations was exposed. The results obtained from [2] were extended by broading formula (1) to the multinomial instances, by showing that those results can be applied suitably to the theoretical cases of study and by the building up of two algorithms which were implemented in two programs in the CAS Maxima. What will be remaining for a subsequent research work will be the developing and programming of an algorithm to implement the use of the general formula obtained in [3], whose output would be something very similar to the one presented here. But overall, some algorithm could be raised, targeted at speed of calculations, to see if this method can be at least as fast as the current ones or even faster.

Thanks

I want to thank my boss, Engineer Eloy Cavazos Galindo, for his constant support and advice during the time I've been working in Trefilados Plant, and for giving me the opportunity to further prepare myself by studying for my master's degree.

Notations

$\mathbb{N}_0$	Union of the set of natural numbers with the zero element
$\hat{n}, \hat{k}$	Entries of binomial or multinomial coefficients
δ_ϕ	Represents the ϕ-th alphabetic character
$\cdots$	Denotes that more summations follow a sequence
$\bullet, \bullet\ldots\bullet$	Denote a product of a summation sequence and a sequence of those products, respectively
$i, j, k, \ldots, \delta_\phi, \ldots$	Indexes of summations
$i^*, j^*, k^*, \ldots, \delta_\phi^*, \ldots$	Same as above but within (*) superscript to contradistinguish from the above
$a \Vert b$	Logic operator OR; it is equivalent to $a \vee b$ (used this way due to space reasons)
$\cdots\sum\sum\sum$	Summation sequence (two or more nested operators)
$\sum_{i=1_c}^{a\Vert b} i_{i,j,k,\ldots}$	Sum of the first a or b numbers. The subindexes $i, j, k, \ldots$ indicate the different indexes of summations corresponding to the sequence it belongs to, where $c = 0, 1, 2, \ldots$ denotes the actual number of a summation sequence
$\binom{a}{b}$	Binomial coefficient
$\left(x_{\{b\}} + y_{\{b\}}\right)^a\Big\vert_{b=\phi}$	Sum of the ϕ-th x plus the ϕ-th y elements, raised to the a-power
$\mathscr{L}$	The generated of a set
$\left[\sum_{i=1}^{1} i_i\right]_{j,k,\ldots}$	Sum of one to one; it equals the unity, written this way to set a starting point and give logic continuity to a sum sequence
∂	The order of a polynomial
⤸	Indicates a sequence of summations continues on the next row

Appendix A: a proof for formula (2)

Here a proof for formula (2) is provided; let us proceed by induction on s. For s amount of summands, its expansion by (2) would be given by the following equation:

$$(x_1+x_2+\cdots+x_s)^n=\prod_{f=1}^{s-1}\left[\sum_{\delta_f^*=0}^{\delta_{f-1}^*}\left(\left\{\sum_{\delta_{\delta_{f-1}^*}=1}^{1}{}_{\delta_{f-1}^*}\cdots\sum_{\delta_{\delta_f^*}=1}^{\left(\delta_{f-1}^*-\delta_f^*\right)+1\|\delta_{\delta_f^*+1}}{}_{\delta_f^*}\cdots\sum_{k=1}^{\left(\delta_{f-1}^*-\delta_f^*\right)+1\|l}{}_2\sum_{j=1}^{\left(\delta_{f-1}^*-\delta_f^*\right)+1\|k}{}_1\left[\sum_{i=1}^{1}{}_0 i_i\right]_{j,k,..,\delta_{\delta_f^*},\dots,\delta_{\delta_{f-1}^*}}\right\}\right)x_f^{\delta_{f-1}^*-\delta_f^*}\right]\bullet\left(x_s^{\delta_{s-1}^*}\right) \tag{36}$$

Fix s = 2 on (6); then the following is obtained:

$$(x_1+x_2)^n=\prod_{f=1}^{1}\left[\sum_{\delta_f^*=0}^{\delta_{f-1}^*}\left(\left\{\sum_{\delta_{\delta_{f-1}^*}=1}^{1}{}_{\delta_{f-1}^*}\cdots\sum_{\delta_{\delta_f^*}=1}^{\left(\delta_{f-1}^*-\delta_f^*\right)+1\|\delta_{\delta_f^*+1}}{}_{\delta_f^*}\cdots\sum_{k=1}^{\left(\delta_{f-1}^*-\delta_f^*\right)+1\|l}{}_2\sum_{j=1}^{\left(\delta_{f-1}^*-\delta_f^*\right)+1\|k}{}_1\left[\sum_{i=1}^{1}{}_0 i_i\right]_{j,k,..,\delta_{\delta_f^*},\dots,\delta_{\delta_{f-1}^*}}\right\}\right)x_f^{\delta_{f-1}^*-\delta_f^*}\right]\bullet\left(x_s^{\delta_{s-1}^*}\right) \tag{37}$$

$$=\sum_{i=0}^{n}\binom{n}{i}x_1^{n-i}x_2^i$$

that actually stands for the binomial expansion for 2−summands; then it is correct. Now, fix $s=k$ on (6) and assume by hypothesis that it is correct.

$$(x_1+x_2+\cdots+x_k)^n=\prod_{f=1}^{k-1}\left[\sum_{\delta_f^*=0}^{\delta_{f-1}^*}\left(\left(\begin{matrix}\delta_{f-1}^*\\ \delta_f^*\end{matrix}\right)\right)x_f^{\delta_{f-1}^*-\delta_f^*}\right]\bullet\left(x_k^{\delta_{k-1}^*}\right) \tag{38}$$

If formula (2) is correct, it must be likewise valid for $s=k+1$. Following up, the attempt is to prove that

$$\prod_{f=1}^{(k+1)-1}\left[\sum_{\delta_f^*=0}^{\delta_{f-1}^*}\left(\left(\begin{matrix}\delta_{f-1}^*\\ \delta_f^*\end{matrix}\right)\right)x_f^{\delta_{f-1}^*-\delta_f^*}\right]\bullet\left(x_{k+1}^{\delta_{(k+1)-1}^*}\right)=(x_1+x_2+\cdots+x_k+x_{k+1})^n \tag{39}$$

First, it will be expanded to the left side of the above equality:

$$\sum_{\delta_1^*=0}^{\delta_0^*}\left(\left\{\sum_{\delta_{\delta_0^*}=1}^{1}{}_{\delta_0^*}\cdots\sum_{\delta_{\delta_1^*}=1}^{\left(\delta_0^*-\delta_1^*\right)+1\|\delta_{\delta_1^*+1}}{}_{\delta_1^*}\cdots\sum_{k=1}^{\left(\delta_0^*-\delta_1^*\right)+1\|l}{}_2\sum_{j=1}^{\left(\delta_0^*-\delta_1^*\right)+1\|k}{}_1\left[\sum_{i=1}^{1}{}_0 i_i\right]_{j,k,..,\delta_{\delta_1^*},\dots,\delta_{\delta_0^*}}\right\}\gamma_1^{\delta_0^*-\delta_1^*}\bullet\right.$$

$$\sum_{\delta_2^*=0}^{\delta_1^*}\left(\left\{\sum_{\delta_{\delta_1^*}=1}^{1}{}_{\delta_1^*}\cdots\sum_{\delta_{\delta_2^*}=1}^{\left(\delta_1^*-\delta_2^*\right)+1\|\delta_{\delta_2^*+1}}{}_{\delta_2^*}\cdots\sum_{k=1}^{\left(\delta_1^*-\delta_2^*\right)+1\|l}{}_2\sum_{j=1}^{\left(\delta_1^*-\delta_2^*\right)+1\|k}{}_1\left[\sum_{i=1}^{1}{}_0 i_i\right]_{j,k,..,\delta_{\delta_2^*},\dots,\delta_{\delta_1^*}}\right\}\gamma_2^{\delta_1^*-\delta_2^*}\bullet\dots\right.$$

$$\dots\bullet\bullet\bullet\dots\bullet\bullet\bullet\dots\bullet\bullet\bullet\dots\bullet\bullet\bullet\dots$$

$$\dots\bullet\sum_{\delta_k^*=0}^{\delta_{k-1}^*}\left(\left\{\sum_{\delta_{\delta_{k-1}^*}=1}^{1}{}_{\delta_{k-1}^*}\cdots\sum_{\delta_{\delta_k^*}=1}^{\left(\delta_{k-1}^*-\delta_k^*\right)+1\|\delta_{\delta_k^*+1}}{}_{\delta_k^*}\cdots\sum_{k=1}^{\left(\delta_{k-1}^*-\delta_k^*\right)+1\|l}{}_2\right.\right. \tag{40}$$

$$\sum_{j=1}^{(\delta_{k-1}^*-\delta_k^*)+1\|k}{}_{1}\left[\sum_{i=1}^{1}{}_{0}\, i_i\right]_{j,k,..,\delta_{\delta_k^*},\,\dots,\,\delta_{\delta_{k-1}^*}}\Bigg\}\gamma_{k-1}^{\delta_{k-1}^*-\delta_k^*}\,\varphi_k^{\delta_{k-1}^*}\Big)\cdot\cdot\Big)\cdots\Big)\cdots\Big)$$

$\Rightarrow$

$$=\sum_{\delta_1^*=0}^{\delta_0^*}\left(\left\{\sum_{\delta_{\delta_0^*}=1}^{1}{}_{\delta_0^*}\cdots\sum_{\delta_{\delta_1^*}=1}^{(\delta_0^*-\delta_1^*)+1\|\delta_{\delta_1^*}+1}{}_{\delta_1^*}\cdots\sum_{k=1}^{(\delta_0^*-\delta_1^*)+1\|l}{}_{2}\ \sum_{j=1}^{(\delta_0^*-\delta_1^*)+1\|k}{}_{1}\left[\sum_{i=1}^{1}{}_{0}\, i_i\right]_{j,k,..,\delta_{\delta_1^*},\,\dots,\,\delta_{\delta_0^*}}\right\}\gamma_1^{\delta_0^*-\delta_1^*}\bullet\right.$$

$$\sum_{\delta_2^*=0}^{\delta_1^*}\left(\left\{\sum_{\delta_{\delta_1^*}=1}^{1}{}_{\delta_1^*}\cdots\sum_{\delta_{\delta_2^*}=1}^{(\delta_1^*-\delta_2^*)+1\|\delta_{\delta_2^*}+1}{}_{\delta_2^*}\cdots\sum_{k=1}^{(\delta_1^*-\delta_2^*)+1\|l}{}_{2}\ \sum_{j=1}^{(\delta_1^*-\delta_2^*)+1\|k}{}_{1}\left[\sum_{i=1}^{1}{}_{0}\, i_i\right]_{j,k,..,\delta_{\delta_2^*},\,\dots,\,\delta_{\delta_1^*}}\right\}\gamma_2^{\delta_1^*-\delta_2^*}\bullet\cdots\right.$$

$$\cdots\ \bull\bullet\bullet\ \dots\bullet\bullet\bullet\ \dots\ \bullet\bullet\bullet\ \dots\bullet\bullet\bullet\ \dots$$

$$\cdots\bullet\sum_{\delta_{(k+1)-1}^*=0}^{\delta_{(k+1)-2}^*}\left(\left\{\sum_{\delta_{\delta_{(k+1)-2}^*}=1}^{1}{}_{\delta_{(k+1)-2}^*}\cdots\sum_{\delta_{\delta_{(k+1)-1}^*}=1}^{\left(\delta_{(k+1)-2}^*-\delta_{(k+1)-1}^*\right)+1\|\delta_{\delta_{(k+1)-1}^*}+1}{}_{\delta_{(k+1)-1}^*}\cdots\sum_{k=1}^{\left(\delta_{(k+1)-2}^*-\delta_{(k+1)-1}^*\right)+1\|l}{}_{2}\right.\right.$$

$$\sum_{j=1}^{\left(\delta_{(k+1)-2}^*-\delta_{(k+1)-1}^*\right)+1\|k}{}_{1}\left[\sum_{i=1}^{1}{}_{0}\, i_i\right]_{j,k,..,\delta_{\delta_{(k+1)-1}^*},\,\dots,\,\delta_{\delta_{(k+1)-2}^*}}\Bigg\}\gamma_{(k+1)-2}^{\delta_{(k+1)-2}^*-\delta_{(k+1)-1}^*}\varphi_{(k+1)-1}^{\delta_{(k+1)-2}^*}\Big)\cdot\cdot\Bigg)\cdots\Bigg)\cdots\Bigg) \tag{41}$$

$$=\prod_{f=1}^{(k+1)-1}\left[\sum_{\delta_f^*=0}^{\delta_{f-1}^*}\left(\left(\left\{\sum_{\delta_{\delta_{f-1}^*}=1}^{1}{}_{\delta_{f-1}^*}\cdots\sum_{\delta_{\delta_f^*}=1}^{\left(\delta_{f-1}^*-\delta_f^*\right)+1\|\delta_{\delta_f^*}+1}{}_{\delta_f^*}\cdots\sum_{k=1}^{\left(\delta_{f-1}^*-\delta_f^*\right)+1\|l1}{}_{2}\right.\right.\right.\right.$$

$$\left.\left.\left.\left.\sum_{j=1}^{\left(\delta_{f-1}^*-\delta_f^*\right)+1\|k}{}_{1}\left[\sum_{i=1}^{1}{}_{0}\, i_i\right]_{j,k,..,\delta_{\delta_f^*},\,\dots,\,\delta_{\delta_{f-1}^*}}\right\}\right)\right)x_f^{\delta_{f-1}^*-\delta_f^*}\right]\bullet\left(x_{k+1}^{\delta_{(k+1)-1}^*}\right)$$

$$=\prod_{f=1}^{(k+1)-1}\left[\sum_{\delta_f^*=0}^{\delta_{f-1}^*}\left(\left(\begin{matrix}\delta_{f-1}^*\\ \delta_f^*\end{matrix}\right)\right)x_f^{\delta_{f-1}^*-\delta_f^*}\right]\bullet\left(x_{k+1}^{\delta_{(k+1)-1}^*}\right)$$

$$=(x_1+x_2+\cdots+x_k+x_{k+1})^n$$

(where it was substituted with $(x_1+x_2+\cdots+x_k)^n$ according to the hypothesis). This result confirms the hypothesis and the validity for (2).

Appendix B: Table

Next on **Table 1**, the symbolic representations for the binomial coefficients are presented. The value actually comes from the output of the program *coef.mc* provided in Section 3.2, after being evaluated with the corresponding values of k and n. This table is shown below.

DOI: http://dx.doi.org/10.5772/intechopen.91422

n	$\binom{n}{1}$	$\binom{n}{2}$	$\binom{n}{3}$	$\binom{n}{4}$	$\binom{n}{5}$	$\binom{n}{6}$	$\binom{n}{7}$	$\binom{n}{8}$	$\binom{n}{9}$
1	$\sum_{i=1}^{1}[1]$								
2	$\sum_{i=1}^{1}[1]$	$\sum_{j=1}^{1}\left[\sum_{i=1}^{j}1\right]$							
3	$\sum_{i=1}^{1}[1]$	$\sum_{j=1}^{2}\left[\sum_{i=1}^{j}1\right]$	$\sum_{k=1}^{1}\sum_{j=1}^{k}\left[\sum_{i=1}^{j}1\right]$						
4	$\sum_{i=1}^{1}[1]$	$\sum_{j=1}^{3}\left[\sum_{i=1}^{j}1\right]$	$\sum_{k=1}^{2}\sum_{j=1}^{k}\left[\sum_{i=1}^{j}1\right]$	$\sum_{l=1}^{1}\sum_{k=1}^{l}\sum_{j=1}^{k}\left[\sum_{i=1}^{j}1\right]$					
5	$\sum_{i=1}^{1}[1]$	$\sum_{j=1}^{4}\left[\sum_{i=1}^{j}1\right]$	$\sum_{k=1}^{3}\sum_{j=1}^{k}\left[\sum_{i=1}^{j}1\right]$	$\sum_{l=1}^{2}\sum_{k=1}^{l}\sum_{j=1}^{k}\left[\sum_{i=1}^{j}1\right]$	$\sum_{m=1}^{1}\sum_{l=1}^{m}\sum_{k=1}^{l}\sum_{j=1}^{k}\left[\sum_{i=1}^{j}1\right]$				
6	$\sum_{i=1}^{1}[1]$	$\sum_{j=1}^{5}\left[\sum_{i=1}^{j}1\right]$	$\sum_{k=1}^{4}\sum_{j=1}^{k}\left[\sum_{i=1}^{j}1\right]$	$\sum_{l=1}^{3}\sum_{k=1}^{l}\sum_{j=1}^{k}\left[\sum_{i=1}^{j}1\right]$	$\sum_{m=1}^{2}\sum_{l=1}^{m}\sum_{k=1}^{l}\sum_{j=1}^{k}\left[\sum_{i=1}^{j}1\right]$	$\sum_{n=1}^{1}\sum_{m=1}^{n}\sum_{l=1}^{m}\sum_{k=1}^{l}\sum_{j=1}^{k}\left[\sum_{i=1}^{j}1\right]$			
7	$\sum_{i=1}^{1}[1]$	$\sum_{j=1}^{6}\left[\sum_{i=1}^{j}1\right]$	$\sum_{k=1}^{5}\sum_{j=1}^{k}\left[\sum_{i=1}^{j}1\right]$	$\sum_{l=1}^{4}\sum_{k=1}^{l}\sum_{j=1}^{k}\left[\sum_{i=1}^{j}1\right]$	$\sum_{m=1}^{3}\sum_{l=1}^{m}\sum_{k=1}^{l}\sum_{j=1}^{k}\left[\sum_{i=1}^{j}1\right]$	$\sum_{n=1}^{2}\sum_{m=1}^{n}\sum_{l=1}^{m}\sum_{k=1}^{l}\sum_{j=1}^{k}\left[\sum_{i=1}^{j}1\right]$	$\sum_{o=1}^{1}\sum_{n=1}^{o}\sum_{m=1}^{n}\sum_{l=1}^{m}\sum_{k=1}^{l}\sum_{j=1}^{k}\left[\sum_{i=1}^{j}1\right]$		
8	$\sum_{i=1}^{1}[1]$	$\sum_{j=1}^{7}\left[\sum_{i=1}^{j}1\right]$	$\sum_{k=1}^{6}\sum_{j=1}^{k}\left[\sum_{i=1}^{j}1\right]$	$\sum_{l=1}^{5}\sum_{k=1}^{l}\sum_{j=1}^{k}\left[\sum_{i=1}^{j}1\right]$	$\sum_{m=1}^{4}\sum_{l=1}^{m}\sum_{k=1}^{l}\sum_{j=1}^{k}\left[\sum_{i=1}^{j}1\right]$	$\sum_{n=1}^{3}\sum_{m=1}^{n}\sum_{l=1}^{m}\sum_{k=1}^{l}\sum_{j=1}^{k}\left[\sum_{i=1}^{j}1\right]$	$\sum_{o=1}^{2}\sum_{n=1}^{o}\sum_{m=1}^{n}\sum_{l=1}^{m}\sum_{k=1}^{l}\sum_{j=1}^{k}\left[\sum_{i=1}^{j}1\right]$	$\sum_{p=1}^{1}\sum_{o=1}^{p}\sum_{n=1}^{o}\sum_{m=1}^{n}\sum_{l=1}^{m}\sum_{k=1}^{l}\sum_{j=1}^{k}\left[\sum_{i=1}^{j}1\right]$	
9	$\sum_{i=1}^{1}[1]$	$\sum_{j=1}^{8}\left[\sum_{i=1}^{j}1\right]$	$\sum_{k=1}^{7}\sum_{j=1}^{k}\left[\sum_{i=1}^{j}1\right]$	$\sum_{l=1}^{6}\sum_{k=1}^{l}\sum_{j=1}^{k}\left[\sum_{i=1}^{j}1\right]$	$\sum_{m=1}^{5}\sum_{l=1}^{m}\sum_{k=1}^{l}\sum_{j=1}^{k}\left[\sum_{i=1}^{j}1\right]$	$\sum_{n=1}^{4}\sum_{m=1}^{n}\sum_{l=1}^{m}\sum_{k=1}^{l}\sum_{j=1}^{k}\left[\sum_{i=1}^{j}1\right]$	$\sum_{o=1}^{3}\sum_{n=1}^{o}\sum_{m=1}^{n}\sum_{l=1}^{m}\sum_{k=1}^{l}\sum_{j=1}^{k}\left[\sum_{i=1}^{j}1\right]$	$\sum_{p=1}^{2}\sum_{o=1}^{p}\sum_{n=1}^{o}\sum_{m=1}^{n}\sum_{l=1}^{m}\sum_{k=1}^{l}\sum_{j=1}^{k}\left[\sum_{i=1}^{j}1\right]$	$\sum_{q=1}^{1}\sum_{p=1}^{q}\sum_{o=1}^{p}\sum_{n=1}^{o}\sum_{m=1}^{n}\sum_{l=1}^{m}\sum_{k=1}^{l}\sum_{j=1}^{k}\left[\sum_{i=1}^{j}1\right]$

Table 1.
Layout pattern of the representation for the first nine binomial coefficients.

Author details

José Alfredo Sánchez de León[1,2]

1 Villacero Trefilados, San Nicolás de los Garza, Nuevo León, México

2 Universidad Autónoma de Nuevo León, San Nicolás de los Garza, Nuevo León, México

*Address all correspondence to: jose.sanchez@villacero.com

References

[1] de León JAS. A model for calculation of a productivity bonus. EUREKA. 2017; **42**:10-27. ISSN: 1405-5090

[2] de León JAS. Calculation of binomial and multinomial coefficients. Mathematics in Computer Science. 2019;**13**:403-415. DOI: 10.1007/s11786-019-00403-w

[3] Rotman J. An Introduction to the Theory of Groups. 4th ed. New York: Springer; 1995. p. 515

[4] Herstein IN. Topics In Algebra. 2nd ed. New York: Wiley; 1975. p. 388

[5] Munkres J. Topology. 2nd ed. Essex: Pearson; 2014. p. 503

[6] Mishra B. Algorithmic Algebra. New York: Springer; 1993. p. 416

[7] Dummit DS, Foote RM. Abstract Algebra. 3rd ed. New York: Wiley; 2004. p. 932

[8] Hernández F. Teoría de Conjuntos. Ciudad de México: Sociedad Matemática Mexicana; 1998. p. 342

[9] Rudin W. Principles of Mathematical Analysis. 3rd ed. New York: McGraw-Hill; 1976. p. 342

[10] Graham RL, Knuth DE, Patashnick O. Concrete Mathematics. 2nd ed. Boston: Addison-Wesley; 1994

[11] Stanley RP. Enumerative Combinatorics. 2nd ed. Vol. 1. Cambridge: Cambridge University Press; 2012

[12] Rusell M. Combinatorics. 2nd ed. Hoboken: Wiley; 2001. p. 576

[13] Fraleigh JB. A First Course in Abstract Algebra. 7th ed. Pearson: Essex; 2002. p. 544

[14] Maxima. Computer Algebra System [Internet]. 1998. Available at: https://maxima.sourceforge.net/ [Accessed: 10 January 2020]

[15] Common LISP. Object-Oriented Programming Language [Internet]. 1984. Available at: https://lisp-lang.org/ [Accessed: 10 January 2020]

[16] Macsyma. Computer Algebra System [Internet]. 1982. Available at: https://swmath.org/software/1209 [Accessed: 10 January 2020]

[17] Roanes-Lozano E, Galán-García JL, Solano-Macías C. Some reflections about the success and impact of the computer algebra system DERIVE with a 10-year time perspective. Mathematics in Computer Science. 2019;**13**:417-431. DOI: 10.1007/s11786-019-00404-9

Chapter 7

How Are Fractal Interpolation Functions Related to Several Contractions?

SongIl Ri and Vasileios Drakopoulos

Abstract

This chapter provides an overview of several types of fractal interpolation functions that are often studied by many researchers and includes some of the latest research made by the authors. Furthermore, it focuses on the connections between fractal interpolation functions resulting from Banach contractions as well as those resulting from Rakotch contractions. Our aim is to give theoretical and practical significance for the generation of fractal (graph of) functions in two and three dimensions for interpolation purposes that are not necessarily associated with Banach contractions.

Keywords: attractor, contraction, fixed point, iterated function system, fractal interpolation

1. Introduction

Interpolation is a method of constructing new data points within the range of a discrete set of known data points or the process of estimating the value of a function at a point from its values at nearby points. Although a large number of interpolation schemes are available in the mathematical field of numerical analysis, the majority of these conventional interpolation methods produce *interpolants*, i.e., functions used to generate interpolation, that are differentiable a number of times except possibly at a finite set of points. Taking into account that the *smoothness* of a function is a property measured by the number of continuous derivatives it has over some domain, the aforementioned interpolants are considered smooth.

On the other hand, many real-world and experimental signals are intricate and rarely show a sensation of smoothness in their traces. Consequently, to model these signals, we require interpolants that are nondifferentiable in dense sets of points in the domain. To address this issue, interpolation by fractal (graph of) functions is introduced in [1, 2], which is based on the theory of *iterated function system*. A *fractal interpolation function* can be considered as a continuous function whose graph is the *attractor*, a fractal set, of an appropriately chosen iterated function system. If this graph has a *Hausdorff-Besicovitch dimension* between 1 and 2, the resulting attractor is called *fractal interpolation curved line* or *fractal interpolation curve*. If this graph has a *Hausdorff-Besicovitch dimension* between 2 and 3, the resulting attractor is called *fractal interpolation surface*. Various types of fractal interpolation functions have been constructed, and some significant properties of them, including calculus, dimension, smoothness, stability, perturbation error, etc., have been widely studied [3–5].

Fractal interpolation is an advanced technique for analysis and synthesis of scientific and engineering data, whereas the approximation of natural curves and surfaces in these areas has emerged as an important research field. *Fractal functions* are currently being given considerable attention due to their applications in areas such as Metallurgy, Earth Sciences, Surface Physics, Chemistry and Medical Sciences. In the development of fractal interpolation theory, many researchers have generalised the notion in different ways [6–9]. Two key issues should be addressed in constructing fractal interpolation functions. They regard to ensuring continuity and the existence of the *contractivity*, or *vertical scaling, factors*; see [10, 11]. In [12], nonlinear fractal interpolation surfaces resulting from Rakotch or Geraghty contractions together with some continuity conditions were introduced as well as explicit illustrative examples were given.

The concept of iterated function system was originally introduced as a generalisation of the well-known Banach contraction principle. Since it has become a powerful tool for constructing and analysing fractal interpolation functions, one can use the well-known fixed point results obtained in the fixed point theory in order to construct them in a more general sense. A comparison of various definitions of contractive mappings as well as fixed point theorems that can be used to construct iterated function systems can be found in [13–15]. In [14], the authors proposed some iterated function systems by using various fixed point theorems, but unfortunately, one does not know whether fractal interpolation functions correspond to those may exist or not. As far as we know, the first significant generalisation of Banach's principle was obtained by Rakotch [16] in 1962. Recently, a method to generate nonlinear fractal interpolation functions by using the Rakotch or Geraghty fixed point theorem instead of Banach fixed point theorem was presented in [12, 17, 18].

The aim of our article is to provide the connections between several fractal interpolation functions and the contractions used to generate them; it is organised as follows. In Section 2, we recall the results obtained in construction of fractal interpolation curved lines and fractal interpolation surfaces by using Rakotch contractions (or Geraghty contractions) instead of Banach contractions. In Section 3, we only present the connection between fractal interpolation functions by using the Banach contractions and fractal interpolation functions by using the Rakotch contractions because in the case of Geraghty contractions, the existence of fractal interpolation curved lines and fractal interpolation surfaces is similar to the case of Rakotch contractions.

2. Preliminaries

Let (X,ρ) and (Y,σ) be metric spaces. A mapping $T: X \to Y$ is called a *Hölder mapping of exponent* or *order a*, if

$$\sigma(T(x), T(y)) \leq c\,[\rho(x,y)]^a$$

for $x, y \in X$, $a \geq 0$ and for some constant c. Note that, if $a > 1$, the functions are constants. Obviously, $c \geq 0$. The mapping T is called a *Lipschitz mapping*, if a may be taken to be equal to 1. If $c = 1$, T is said to be *nonexpansive*. A Lipschitz function is a *contraction* with *contractivity factor* c, if $c < 1$. We call T *contractive*, if for all $x, y \in X$ and $x \neq y$, we have $\sigma(T(x), T(y)) < \rho(x,y)$. Note that 'contraction $\Rightarrow$ contractive $\Rightarrow$ nonexpansive $\Rightarrow$ Lipschitz'.

An *iterated function system*, or *IFS* for short, is a collection of a complete metric space (X, ρ) together with a finite set of continuous mappings, $f_n : X \to X$, $n = 1, 2, \ldots, N$. It is often convenient to write an IFS formally as $\{X; f_1, f_2, \ldots, f_N\}$ or, somewhat more briefly, as $\{X; f_{1-N}\}$. The associated map of subsets $W : \mathcal{H}(X) \to \mathcal{H}(X)$ is given by:

$$W(E) = \bigcup_{n=1}^{N} f_n(E) \quad \text{for all} \quad E \in \mathcal{H}(X),$$

where $\mathcal{H}(X)$ is the metric space of all nonempty, compact subsets of X with respect to some metric, e.g., the Hausdorff metric. The map W is called the *Hutchinson operator* or the *collage map* to alert us to the fact that $W(E)$ is formed as a union or 'collage' of sets.

If w_n are contractions with corresponding contractivity factors s_n for $n = 1, 2, \ldots, N$, the IFS is termed *hyperbolic* and the map W itself is then a contraction with contractivity factor $s = \max\{s_1, s_2, \ldots, s_N\}$ ([2], Theorem 7.1, p. 81). In what follows, we abbreviate by f^k the k-fold composition $f \circ f \circ \cdots \circ f$.

Definition 2.1. *Let X be a set. A* self-map *on X or a* transformation *is a mapping from X to itself.*

i. *A self-map f on a metric space (X, ρ) is called a φ-*contraction*, if there exists a function $\varphi : (0, +\infty) \to (0, +\infty)$ with $\phi(0) = 0$ and $\phi(t) < t$ for all $t > 0$ such that for all $x, y \in X$, $\rho(f(x), f(y)) \le \varphi(\rho(x, y))$.*

ii. *We say that f is a* Rakotch contraction, *iff f is a φ-contraction such that for any $t > 0$, $\alpha(t) := \frac{\varphi(t)}{t} < 1$ and the function $(0, +\infty) \ni t \to \frac{\varphi(t)}{t}$ is nonincreasing.*

iii. *If f is a φ-contraction for some function $\varphi : (0, +\infty) \to (0, +\infty)$ such that for any $t > 0$, $\alpha(t) := \frac{\varphi(t)}{t} < 1$ and the function $(0, +\infty) \ni t \to \frac{\varphi(t)}{t}$ is nonincreasing (or nondecreasing, or continuous), then we call such a function a* Geraghty contraction.

From [14], we have the following.

Theorem 2.1. *Let X be a complete metric space and $\{X; f_{1-N}\}$ be an IFS consisting of Rakotch or Geraghty contractions. Then there is a unique nonempty compact set $K \in \mathcal{H}(X)$ such that*

$$K = \bigcup_{n=1}^{N} f_n(K).$$

2.1 Fractal interpolation in $\mathbb{R}$

Let N be a positive integer greater than 1 and $I = [x_0, x_N] \subset \mathbb{R}$. Let a set of *interpolation points* $\{(x_i, y_i) \in I \times \mathbb{R} : i = 0, 1, \ldots, N\}$ be given, where $x_0 < x_1 < \cdots < x_N$ and $y_0, y_1, \ldots, y_N \in \mathbb{R}$. Set $I_n = [x_{n-1}, x_n] \subset I$ and define, for all $n = 1, 2, \ldots, N$, contractive homeomorphisms $L_n : I \to I_n$ by

$$L_n(x) := a_n x + b_n,$$

where the real numbers a_n, b_n are chosen to ensure that $L_n(I) = I_n$.

Let $\varphi : (0, +\infty) \to (0, +\infty)$ be a nondecreasing continuous function such that for any $t > 0$, $\alpha(t) := \frac{\varphi(t)}{t} < 1$ and the function $(0, +\infty) \ni t \to \frac{\varphi(t)}{t}$ is nonincreasing. Let $d_n : I \to \mathbb{R}$ be a continuously differentiable function such that

$$\max_{x\in I} |d_n(x)| \leq 1.$$

Now, consider an IFS of the form $\{I \times \mathbb{R}; w_n, n = 1, 2, \ldots, N\}$ in which the maps are nonlinear transformations of the special structure

$$w_n\begin{pmatrix} x \\ y \end{pmatrix} = \begin{pmatrix} L_n(x) \\ F_n(x,y) \end{pmatrix} = \begin{pmatrix} a_n x + b_n \\ c_n x + d_n(x) s_n(y) + e_n \end{pmatrix},$$

where the transformations are constrained by the data according to

$$w_n\begin{pmatrix} x_0 \\ y_0 \end{pmatrix} = \begin{pmatrix} x_{n-1} \\ y_{n-1} \end{pmatrix}, \quad w_n\begin{pmatrix} x_N \\ y_N \end{pmatrix} = \begin{pmatrix} x_n \\ y_n \end{pmatrix}$$

for $n = 1, 2, \ldots, N$, and s_n are some Rakotch or Geraghty contractions.

Let us denote by $C(D)$ the linear space of all real-valued continuous functions defined on D, i.e., $C(D) = \{ f : D \to \mathbb{R} \mid f \text{ continuous}\}$. Let $C^*(I) \subset C(I)$ denote the set of continuous functions $f : I \to \mathbb{R}$ such that $f(x_0) = y_0$ and $f(x_N) = y_N$, that is,

$$C^*(I) := \{ f \in C(I) : f(x_0) = y_0, f(x_N) = y_N\}.$$

Let $C^{**}(I) \subset C^*(I) \subset C(I)$ be the set of continuous functions that pass through the given data points $\{(x_i, y_i) \in I \times \mathbb{R} : i = 0, 1, \ldots, N\}$, that is,

$$C^{**}(I) := \{ f \in C^*(I) : f(x_i) = y_i, i = 0, 1, \ldots, N\}.$$

Define a metric $d_{C(I)}$ on the space $C(I)$ by

$$d_{C(I)}(g,h) := \max_{x\in[x_0,x_N]} |g(x) - h(x)|$$

for all $g, h \in C(I)$. Define a mapping $T\colon C^*(I) \to C(I)$ for all $f \in C^*(I)$ by

$$\begin{aligned} Tf(x) &:= F_n\left(L_n^{-1}(x), f\left(L_n^{-1}(x)\right)\right) \\ &= c_n L_n^{-1}(x) + d_n\left(L_n^{-1}(x)\right) s_n\left(f\left(L_n^{-1}(x)\right)\right) + e_n \end{aligned}$$

for $x \in [x_{n-1}, x_n]$ and $n = 1, 2, \ldots, N$. From [17], we have the following.

Theorem 2.2. *Let $\{I \times \mathbb{R}; w_n, n = 1, 2, \ldots, N\}$ denote the IFS defined above. Let each s_n be a bounded Rakotch or Geraghty contraction. Then,*

i. *there is a unique continuous function $f : I \to \mathbb{R}$ which is a fixed point of T;*

ii. *$f(x_i) = y_i$ for all $i = 0, 1, \ldots, N$;*

iii. *if $G \subset I \times \mathbb{R}$ is the graph of f, then*

$$G = \bigcup_{n=1}^{N} w_n(G).$$

An extremely explicit simple example is the following; cf. [12].

Example 1. *Let $\varphi(t) := \frac{t}{1+t}$ for $t \in (0, +\infty)$. Let a set of data $\{(x_i, y_i) : i = 0, 1, \ldots, N\}$ be given, where $0 = x_0 < x_1 < \ldots < x_N = 1$ and $y_i \in [0, 1]$ for all $i = 0, 1, \ldots, N$. Let for all $n = 1, 2, \ldots, N$, $d_n(x) := x^n$. Let for $y \in [0, +\infty)$ and*

$n = 1, 2, \ldots, N$, $s_n(y) := \frac{y}{1+ny}$. *That is, each s_n is a Rakotch contraction (with the same function φ) that is not a Banach contraction on $[0, +\infty)$. Let for all $n = 1, 2, \ldots, N$,*

$$w_n(x, y) := (a_n x + b_n, c_n x + d_n(x) s_n(y) + e_n),$$

where

$$a_n = x_n - x_{n-1}, \quad b_n = x_{n-1},$$

$$c_n = y_n - y_{n-1}, \quad e_n = y_{n-1}.$$

Then, there exists a continuous function $f : [0, 1] \to \mathbb{R}$ that interpolates the given points $\{(x_i, y_i) : i = 0, 1, \ldots, N\}$. Moreover, the graph G of f is invariant with respect to $\{[0, 1] \times \mathbb{R}; w_1, w_2, \ldots, w_N\}$, i.e.,

$$G = \bigcup_{n=1}^{N} w_n(G).$$

2.2 Fractal interpolation in $\mathbb{R}^2$

Let M, N be two positive integers greater than 1. Let us represent the given set of *interpolation points* as $\left\{\left(x_i, y_j, z_{i,j}\right) \in K : i = 0, 1, \ldots, M; j = 0, 1, \ldots, N\right\}$, where $x_0 < x_1 < \cdots < x_M$, $y_0 < y_1 < \cdots < y_N$ and $z_{i,j} \in [a, b]$ for all $i = 0, 1, \ldots, M$ and $j = 0, 1, \ldots, N$. Set $I = [x_0, x_M] \subset \mathbb{R}$ and $J = [y_0, y_N] \subset \mathbb{R}$. Throughout this section, we will work in the complete metric space $K = D \times \mathbb{R}$, where $D = I \times J$, with respect to the Euclidean, or to some other equivalent, metric.

Set $I_m = [x_{m-1}, x_m]$, $J_n = [y_{n-1}, y_n]$, $D_{m,n} = I_m \times J_n$ and let $u_m : I \to I_m$, $v_n : J \to J_n$, $L_{m,n} : D \to D_{m,n}$ be defined for $m = 1, 2, \ldots, M$ and $n = 1, 2, \ldots, N$, by

$$L_{m,n}(x, y) = (u_m(x), v_n(y)) = (a_m x + b_m, c_n y + d_n).$$

Thus, for $m = 1, 2, \ldots, M$ and $n = 1, 2, \ldots, N$,

$$a_m = \frac{x_m - x_{m-1}}{x_M - x_0}, \quad b_m = x_{m-1} - \frac{x_m - x_{m-1}}{x_M - x_0} x_0,$$

$$c_n = \frac{y_n - y_{n-1}}{y_N - y_0}, \quad d_n = y_{n-1} - \frac{y_n - y_{n-1}}{y_N - y_0} y_0.$$

Furthermore, for $m = 1, 2, \ldots, M$ and $n = 1, 2, \ldots, N$, let mappings $F_{m,n} : K \to \mathbb{R}$ be continuous with respect to each variable. We consider an IFS of the form $\{K; w_{m,n}, m = 1, 2, \ldots, M; n = 1, 2, \ldots, N\}$ in which maps $w_{m,n} : D \times \mathbb{R} \to D_{m,n} \times \mathbb{R}$ are transformations of the special structure

$$w_{m,n}(x, y, z) := (L_{m,n}(x, y), F_{m,n}(x, y, z)),$$

where the transformations are constrained by the data according to

$$w_{m,n}\begin{pmatrix} x_0 \\ y_0 \\ z_{0,0} \end{pmatrix} = \begin{pmatrix} x_{m-1} \\ y_{n-1} \\ z_{m-1,n-1} \end{pmatrix}, \quad w_{m,n}\begin{pmatrix} x_0 \\ y_N \\ z_{0,N} \end{pmatrix} = \begin{pmatrix} x_{m-1} \\ y_n \\ z_{m-1,n} \end{pmatrix},$$

$$w_{m,n}\begin{pmatrix} x_M \\ y_0 \\ z_{M,0} \end{pmatrix} = \begin{pmatrix} x_m \\ y_{n-1} \\ z_{m,n-1} \end{pmatrix}, \quad w_{m,n}\begin{pmatrix} x_M \\ y_N \\ z_{M,N} \end{pmatrix} = \begin{pmatrix} x_m \\ y_n \\ z_{m,n} \end{pmatrix}$$

for $m = 1, 2, \dots, M$ and $n = 1, 2, \dots, N$.

Let $B(D)$ denote the set of bounded functions $f : D \to \mathbb{R}$ and

$$B^*(D) = \{f \in B(D) : f(x_0, y_0) = z_{0,0},\ f(x_0, y_N) = z_{0,N},$$
$$f(x_M, y_0) = z_{M,0}, f(x_M, y_N) = z_{M,N}\}.$$

Let $B^{**}(D) \subset B^*(D)$ be the set of bounded functions that pass through the given interpolation points $\left\{\left(x_i, y_j, z_{i,j}\right) \in K = D \times [a, b] : i = 0, 1, \dots, M; j = 0, 1, \dots, N\right\}$, that is,

$$B^{**}(D) = \left\{ f \in B^*(D) : f\left(x_i, y_j\right) = z_{i,j}, i = 0, 1, \dots, M; j = 0, 1, \dots, N\right\}.$$

Define an operator $T : B^*(D) \to B(D)$ for all $f \in B^*(D)$ by

$$Tf(x, y) = F_{m,n}\left(u_m^{-1}(x), v_n^{-1}(y), f\left(u_m^{-1}(x), v_n^{-1}(y)\right)\right)$$

for $(x, y) \in D_{m,n}$, $m = 1, 2, \dots, M$ and $n = 1, 2, \dots, N$. In [18], we see the following.

Theorem 2.3. *Let $\{D \times \mathbb{R}; w_{m,n}, m = 1, 2, \dots, M; n = 1, 2, \dots, N\}$ denote the IFS defined above. Assume that the maps $F_{m,n}$ are Rakotch or Geraghty contractions with respect to the third variable, and uniformly Lipschitz with respect to the first and second variable. Then,*

1. *there is a unique bounded function $f : D \to \mathbb{R}$ which is a fixed point of T;*

2. $f\left(x_i, y_j\right) = z_{i,j}$ *for $i = 0, 1, \dots, M$ and $j = 0, 1, \dots, N$;*

3. *if $G \subset D \times \mathbb{R}$ is the graph of f, then*

$$G = \bigcup_{m=1}^{M} \bigcup_{n=1}^{N} w_{m,n}(G).$$

Let for all $i = 0, 1, \dots, M$ and $j = 0, 1, \dots, N$, $z_{0,j} = z_{i,0} = z_{M,j} = z_{i,N}$ and define

$$F_{m,n}(x, y, z) = e_{m,n}x + f_{m,n}y + g_{m,n}xy + s_{m,n}(z) + h_{m,n},$$

where $s_{m,n}$ are Rakotch or Geraghty contractions. Let

$$C^*(D) = \{f \in C(D) : f(x_0, y_0) = z_{0,0}, f(x_0, y_N) = z_{0,N},$$
$$f(x_M, y_0) = z_{M,0}, f(x_M, y_N) = z_{M,N}\}$$

and

$$C^{**}(D) = \left\{ f \in C^*(D) : f\left(x_i, y_j\right) = z_{i,j}, i = 0, 1, \dots, M; j = 0, 1, \dots, N\right\}.$$

Let $C_0^*(D) \subset C^*(D)$ be the set of continuous functions $f : D \to \mathbb{R}$ such that

$$\begin{aligned} f\left(x_0, (1-\lambda)y_0 + \lambda y_N\right) &= z_{*,*}, \\ f\left(x_M, (1-\lambda)y_0 + \lambda y_N\right) &= z_{*,*}, \\ f\left((1-\lambda)x_0 + \lambda x_M, y_0\right) &= z_{*,*}, \\ f\left((1-\lambda)x_0 + \lambda x_M, y_N\right) &= z_{*,*} \end{aligned}$$

for all $\lambda \in [0,1]$, where for all $i = 0, 1, \ldots, M$ and $j = 0, 1, \ldots, N$,

$$z_{*\,*} := z_{0,j} = z_{i,0} = z_{M,j} = z_{i,N}.$$

Let $C_0^{**}(D) := \left\{ f \in C_0^*(D) : f\left(x_i, y_j\right) = z_{i,j}, i = 0, 1, \ldots, M; j = 0, 1, \ldots, N \right\} \subset C^{**}(D)$. For $f \in C_0^*(D)$, we define $T : C_0^*(D) \to B(D)$ by

$$\begin{aligned} Tf(x,y) &= F_{m,n}\left(u_m^{-1}(x), v_n^{-1}(y), f\left(u_m^{-1}(x), v_n^{-1}(y)\right)\right) \\ &= e_{m,n}u_m^{-1}(x) + f_{m,n}v_n^{-1}(y) + g_{m,n}u_m^{-1}(x)v_n^{-1}(y) \\ &\quad + s_{m,n}\left(f\left(u_m^{-1}(x), v_n^{-1}(y)\right)\right) + h_{m,n} \end{aligned}$$

for $(x,y) \in D_{m,n}$, $m = 1, 2, \ldots, M$ and $n = 1, 2, \ldots, N$.

Corollary 2.1 *(see* [18]*) Let* $\{D \times \mathbb{R}; w_{m,n}, m = 1, 2, \ldots, M; n = 1, 2, \ldots, N\}$ *denote the IFS defined above. Then,*

1. *there is a unique continuous function* $f : D \to \mathbb{R}$ *which is a fixed point of* T*;*

2. $f\left(x_i, y_j\right) = z_{i,j}$ *for all* $i = 0, 1, \ldots, M$ *and* $j = 0, 1, \ldots, N$*;*

3. *if* $G \subset D \times \mathbb{R}$ *is the graph of* f*, then*

$$G = \bigcup_{m=1}^{M} \bigcup_{n=1}^{N} w_{m,n}(G).$$

The most simple example is the following; cf. [12].

Example 2. *Let* $\varphi(t) := \frac{t}{1+t}$ *for* $t \in (0, +\infty)$*. Let a set of data* $\left\{\left(x_i, y_j, z_{i,j}\right) : i = 0, 1, 2; j = 0, 1, 2\right\}$ *be given, where* $0 = x_0 < x_1 < x_2 = 1$, $0 = y_0 < y_1 < y_2 = 1$ *and* $z_{i,j} \in [0,1]$ *for all* $i = 0, 1, 2; j = 0, 1, 2$*. Let for all* $i = 0, 1, 2$ *and* $j = 0, 1, 2$,

$$z_{0,j} = z_{i,0} = z_{2,j} = z_{i,2} = 0.$$

Let for $z \in [0, +\infty)$,

$$\begin{aligned} s_{1,1}(z) &:= \frac{z}{1+z}, \quad s_{1,2}(z) := \frac{z}{1+2z}, \\ s_{2,1}(z) &:= \frac{z}{1+3z}, \quad s_{2,2}(z) := \frac{z}{1+4z}. \end{aligned}$$

Then, $s_{1,1}, s_{1,2}, s_{2,1}, s_{2,2}$ *are Rakotch contractions (with the same function* φ*) that are not Banach contractions on* $[0, +\infty)$*. So, there exists a continuous function* $f : [0,1] \times [0,1] \to \mathbb{R}$ *that interpolates the given data* $\left\{\left(x_i, y_j, z_{i,j}\right) : i = 0, 1, 2; j = 0, 1, 2\right\}$.

Let $d_{m,n} : D \to \mathbb{R}$ be a function such that $\max_{(x,y) \in D} |d_{m,n}(x,y)| \le 1$,

$$d_{m,n}(x_0,y) = d_{m,n}(x_M,y) = d_{m,n}(x,y_0) = d_{m,n}(x,y_N) = 0$$

and for some $L_1, L_2 > 0$,

$$|d_{m,n}(x,y) - d_{m,n}(x',y')| \leq L_1|x - x'| + L_2|y - y'|.$$

Let

$$F_{m,n}(x,y,z) = e_{m,n}x + f_{m,n}y + g_{m,n}xy + d_{m,n}(x,y)s_{m,n}(z) + h_{m,n},$$

where $s_{m,n}$ is a Rakotch or Geraghty contraction. For $f \in C^*(D)$, we define $T : C^*(D) \to B(D)$ by

$$Tf(x,y) = F_{m,n}\left(u_m^{-1}(x), v_n^{-1}(y), f\left(u_m^{-1}(x), v_n^{-1}(y)\right)\right)$$

$$= e_{m,n}u_m^{-1}(x) + f_{m,n}v_n^{-1}(y) + g_{m,n}u_m^{-1}(x)v_n^{-1}(y)$$

$$+ d_{m,n}\left(u_m^{-1}(x), v_n^{-1}(y)\right)s_{m,n}\left(f\left(u_m^{-1}(x), v_n^{-1}(y)\right)\right) + h_{m,n}$$

for $(x,y) \in D_{m,n}$, $m = 1, 2, \ldots, M$ and $n = 1, 2, \ldots, N$.

For the next, see [12] for details.

Corollary 2.2. *Let $\{D \times \mathbb{R}; w_{m,n}, m = 1, 2, \ldots, M; n = 1, 2, \ldots, N\}$ denote the IFS defined above. If each $s_{m,n}$ be a bounded function, then.*

1. *there is a unique continuous function $f : D \to \mathbb{R}$ which is a fixed point of T;*

2. *$f\left(x_i, y_j\right) = z_{i,j}$ for all $i = 0, 1, \ldots, M$ and $j = 0, 1, \ldots, N$;*

3. *if $G \subset D \times \mathbb{R}$ is a graph of f, then*

$$G = \bigcup_{m=1}^{M}\bigcup_{n=1}^{N} w_{m,n}(G).$$

An especially simple example is the following; see [12].

Example 3. *Let $\varphi(t) := \frac{t}{1+t}$ for $t \in (0, +\infty)$. Let a set of data $\left\{\left(x_i, y_j, z_{i,j}\right) : i = 0, 1, 2; j = 0, 1, 2\right\}$ be given, where $0 = x_0 < x_1 < x_2 = 1$, $0 = y_0 < y_1 < y_2 = 1$ and $z_{i,j} \in [0, 1]$ for all $i = 0, 1, 2; j = 0, 1, 2$. Here, a set of data points is not necessarily the case that $z_{0,j} = z_{i,0} = z_{2,j} = z_{i,2}$ for all $i = 0, 1, 2; j = 0, 1, 2$. Let for all $i = 1, 2; j = 1, 2$ and $(x,y) \in [0, 1] \times [0, 1]$,*

$$d_{m,n}(x,y) := 2^{2(m+n)}x^m(1-x)^m y^n(1-y)^n.$$

Let for $z \in [0, +\infty)$,

$$s_{1,1}(z) := \frac{1}{1+z}, s_{1,2}(z) := \frac{z}{1+z},$$

$$s_{2,1}(z) := \frac{z}{1+2z}, s_{2,2}(z) := \frac{z}{1+3z}.$$

Then, $s_{1,1}, s_{1,2}, s_{2,1}, s_{2,2}$ *are Rakotch contractions (with the same function* φ*) that are not Banach contractions on* $[0, +\infty)$*. So, there exists a continuous function* $f : [0,1] \times [0,1] \to \mathbb{R}$ *that interpolates the given data* $\left\{\left(x_i, y_j, z_{i,j}\right) : i = 0,1,2; j = 0,1,2\right\}$.

3. Interconnections between FIFs and contractions

In this section, we only present the interconnections between FIFs resulting from Banach contractions and FIFs resulting from Rakotch contractions because in the case of Geraghty contractions, the existence of FICs and FISs is derived similarly to the case of Rakotch contractions.

Connection 1

1. Each Banach contraction is a Rakotch contraction, since a self-map is a Banach contraction if and only if it is a φ-contraction for a function $\varphi(t) = \alpha t$, for some $0 \leq \alpha < 1$. There exist examples of Rakotch contraction maps that are not Banach contraction maps on $X \subset \mathbb{R}$ with respect to the Euclidean metric (see [13]).

2. The Rakotch's functional condition for convergence of a contractive iteration in a complete metric space can be replaced by an equivalent (or another) functional condition; for instance, a map is a Rakotch contraction if and only if it is a φ-contraction for some nondecreasing function $\varphi : (0, +\infty) \to (0, +\infty)$ such that additionally $\varphi(t) < t$ for $t > 0$ and the map $t \to \frac{\varphi(t)}{t}$ is nonincreasing (see [19]).

Connection 2

1. $\left(C(I), d_{C(I)}\right)$, $\left(C^*(I), d_{C(I)}\right)$ and $\left(C^{**}(I), d_{C(I)}\right)$ are complete metric spaces, where

$$d_{C(I)}(f,g) := \max_{x \in I} |f(x) - g(x)|$$

for all $f, g \in C(I)$ (see [2]).

2. $\left(B(D), d_{B(D)}\right)$, $\left(B^*(D), d_{B(D)}\right)$ and $\left(B^{**}(D), d_{B(D)}\right)$ are complete metric spaces, where

$$d_{B(D)}(f,g) := \sup_{(x,y) \in D} |f(x,y) - g(x,y)|$$

for all $f, g \in B(D)$ [10].

3. $C_0^{**}(D)$, $C_0^*(D)$, $C^{**}(D)$, $C^*(D)$ and $C(D)$ are closed subspaces of $B(D)$ with $C_0^{**}(D) \subset C_0^*(D) \subset C^*(D) \subset C(D) \subset B(D)$ and $C_0^{**}(D) \subset C^{**}(D) \subset C^*(D) \subset C(D) \subset B(D)$, and so they are complete metric spaces.

Connection 3

Let $d_n : I \to \mathbb{R}$ be a continuously differentiable function such that

$$\max_{x \in I} |d_n(x)| \leq 1.$$

Then, by the Differential Mean Value Theorem and the extreme value theorem, we can see that for some $L_{d_n} > 0$,

$$|d_n(x') - d_n(x'')| \leq L_{d_n} |x' - x''|,$$

where $x', x'' \in I$. Hence, d_n is Lipschitz continuous function defined on I satisfying $\max_{x \in I} |d_n(x)| \leq 1$, but the converse is not true in general.

Connection 4

1. The function $d_n(x)s_n(y)$ is a generalisation of the bivariable function $d_n(x)y$ with vertical scaling factors as (continuous) 'contraction functions'. In fact, in the case when $0 < \max_{x \in I} |d_n(x)| < 1$ (see [20], p. 3), obviously,

$$d_n(x)y = \frac{d_n(x)}{\max_{x \in I} |d_n(x)|} \max_{x \in I} |d_n(x)| y.$$

Let $s_n(y) = \max_{x \in I} |d_n(x)| y$ and $d_n^*(x) = \frac{d_n(x)}{\max_{x \in I} |d_n(x)|}$. Then $d_n(x)y = d_n^*(x)s_n(y)$, $\max_{x \in I} |d_n^*(x)| = 1$ and s_n is a Banach (or Rakotch) contraction.

2. The functional condition $\max_{x \in I} |d_n(x)| \leq 1$ is essential in order to show the difference between Banach contractibility of $F_n(\cdot, y)$ and Rakotch contractibility of $F_n(\cdot, y)$; compare with [20]. In fact, since $\varphi(t) < t$ for any $t > 0$,

$$\begin{aligned} |F_n(x, y') - F_n(x, y'')| &= |d_n(x)||s_n(y') - s_n(y'')| \\ &\leq \max_{x \in I} |d_n(x)||s_n(y') - s_n(y'')| \\ &\leq \max_{x \in I} |d_n(x)| \varphi(|y' - y''|) \\ &\leq \max_{x \in I} |d_n(x)||y' - y''|, \end{aligned}$$

where $(x, y'), (x, y'') \in \mathbb{R}^2$. Hence, if $\max_{x \in I} |d_n(x)| < 1$, as can be seen, notwithstanding each s_n is a Rakotch contraction that is not a Banach contraction, each F_n is Banach contraction with respect to the second variable because

$$|F_n(x, y') - F_n(x, y'')| \leq \max_{x \in I} |d_n(x)||y' - y''|.$$

On the other hand, if $\max_{x \in I} |d_n(x)| = 1$, then we conclude that each F_n is Rakotch contraction with respect to the second variable whenever each s_n is a Rakotch contraction because

$$|F_n(x, y') - F_n(x, y'')| \leq \max_{x \in I} |d_n(x)| \varphi(|y' - y''|).$$

3. In Theorem 2.2, for all $(x, y'), (x, y'') \in I \times \mathbb{R}$,

$$|F_n(x, y') - F_n(x, y'')| = |d_n(x)||s_n(y') - s_n(y'')| \le |s_n(y') - s_n(y'')| \le \varphi(|y' - y''|).$$

That is, each $w_n(x, y)$ is chosen so that function $F_n(x, y)$ is Rakotch contraction with respect to the second variable.

4. Even though $s_n : \mathbb{R} \to \mathbb{R}$ are Rakotch contractions, $w_n : I \times \mathbb{R}$ are not in general Rakotch contractions on the metric space $(I \times \mathbb{R}, d_0)$, and thus, the IFSs defined above are not IFSs of [14] (cf. second and third line in p. 215 of [2]).

Connection 5

In the case where the vertical scaling factors are constants, in [1], the existence of affine FIFs by using the Banach fixed point theorem was investigated, whereas in [20], a generalisation of affine FIFs by using vertical scaling factors as (continuous) 'contraction functions' and Banach's fixed point theorem was introduced. Theorem 2.2 gives the existence of fractal interpolation curves by using the Rakotch fixed point theorem and vertical scaling factors as (continuous) 'contraction functions'.

Connection 6

The boundedness of s_n is the essential condition to establish a unique invariant set of an iterated function system. In the fractal interpolation curve with vertical scaling factors as 'contraction function', $0 < \max_{x \in I}|d_n(x)| < 1$ (see [20]). Let $M := \max_{x \in I}|c_n x + f_n|$ and $h \ge \frac{M}{1 - \max_{x \in I}|d_n(x)|}$. Then for all $y \in [-h, h]$,

$$|F_n(x, y)| = |c_n x + d_n(x) y + f_n| \le M + \max_{x \in I}|d_n(x)||y| \le M + \max_{x \in I}|d_n(x)|h \le h.$$

So, for all $(x, y) \in I \times [-h, h]$, we can see that $F_n(x, y) \in [-h, h]$. That is, an IFS of the form $\{I \times [-h, h]; w_{1-N}\}$ has been constructed (cf. [21], p. 1897). Thus $D(s_n) = [-h, h]$ and $s_n(y) := \max_{x \in I}|d_n(x)|y$ is bounded in $D(s_n)$. Hence the boundedness of s_n in $D(s_n)$ is the essential condition to establish a unique invariant set of an IFS (cf. [21], p. 1897).

Connection 7

In view of a φ-contraction, the connections between the coefficients of y variable are obtained as follows:

1. In the affine FIF (cf. [1], p. 308, Example 1), for all $t > 0$,

$$\varphi(t) := \max_{n=1,2,\ldots,N} |d_n| t,$$

where $|d_n| < 1$ for all $i = 1, 2, \ldots, N$.

2. In the FIF with vertical scaling factors as (continuous) 'contraction functions' (cf. [20], p. 3), for all $t > 0$,

$$\varphi(t) := \max_{i=1,2,\ldots,N} \max_{x \in I} |d_n(x)| t,$$

where $d_n(x)$ is Lipschitz function defined on I satisfying $\sup_{x \in I}|d_n(x)| < 1$ for all $n = 1, 2, \ldots, N$.

Connection 8

We refer to f of Theorem 2.2 as a nonlinear FIF. The reason is that the functions F_n take the form

$$F_n(x,y) = c_n x + d_n(x)s_n(y) + e_n,$$

where $\max_{x\in I}|d_n(x)| \le 1$ and each s_n is Rakotch contraction. That is, each F_n, in general, is nonlinear with respect to the second variable (cf. [17]). In fact, in [2] or [20], since $0 < |d_n(x)| \equiv |d_n| < 1$ or $0 < \max_{x\in I}|d_n(x)| < 1$ and

$$d_n(x)y = \frac{d_n(x)}{\max\limits_{x\in I}|d_n(x)|}\max_{x\in I}|d_n(x)|y,$$

we can see that

$$F_n(x,y) = c_n x + d_n(x)y + e_n = c_n x + d_n^*(x)s_n(y) + e_n,$$

where $d_n^*(x) := \frac{d_n(x)}{\max_{x\in I}|d_n(x)|}$ and $s_n(y) := \max_{x\in I}|d_n(x)|y$, and thus, each s_n is a special Banach contraction and linear with respect to the second variable. Obviously, we can say that nonlinear FIFs may have more flexibility and applicability.

Connection 9

1. The well-known FIS in theory and applications is generated by an IFS of the form $\{K, w_{m,n} : m = 1, 2, \ldots, M; n = 1, 2, \ldots, N\}$ under some conditions, where the maps are transformations of the special structure

$$w_{m,n}\begin{pmatrix} x \\ y \\ z \end{pmatrix} = \begin{pmatrix} u_m(x) \\ v_n(y) \\ F_{m,n}(x,y,z) \end{pmatrix} = \begin{pmatrix} a_m x + b_m \\ c_n y + d_n \\ e_{m,n}x + f_{m,n}y + g_{m,n}xy + d_{m,n}(x,y)z + h_{m,n} \end{pmatrix},$$

where $|d_{m,n}(x,y)| < 1$ for all $(x,y) \in D \subset \mathbb{R}^2$. Then for all $(x,y,z), (x,y,z') \in K$,

$$\begin{aligned} |F_{m,n}(x,y,z) - F_{m,n}(x,y,z')| &= |d_{m,n}(x,y)z - d_{m,n}(x,y)z'| \\ &\le \max_{(x,y)\in D}|d_{m,n}(x,y)||z - z'|. \end{aligned}$$

That is, each $w_{m,n}(x,y,z)$ is chosen so that function $F_{m,n}(x,y,z)$ is a Banach contraction with respect to the third variable. So, the existence of bivariable FIFs follows from Banach's fixed point theorem. In fact, in [22], since for all $(x,y) \in D \subset \mathbb{R}^2$, $d_{m,n}(x,y) \equiv s_{m,n}$ and $0 \le |s_{m,n}| < 1$, we can see that each $w_{m,n}(x,y,z)$ is chosen so that function $F_{m,n}(x,y,z)$ is Banach contraction with respect to the third variable. Also in [21], since

$$d_{m,n}(x,y) = \lambda_{m,n}(x - x_0)(x_M - x)(y - y_0)(y_N - y)$$

and

$$|\lambda_{m,n}| < \frac{16}{(x_M - x_0)^2(y_N - y_0)^2},$$

we can see that $\max_{(x,y)\in D}|d_{m,n}(x,y)|<1$, and so each $w_{m,n}(x,y,z)$ is chosen so that function $F_{m,n}(x,y,z)$ is a Banach contraction with respect to the third variable.

2. In Theorem 2.3, for all $(x,y,z),(x,y,z')\in K\subset\mathbb{R}^3$,

$$|F_{m,n}(x,y,z)-F_{m,n}(x,y,z')|=|d_{m,n}(x,y)||s_{m,n}(z)-s_{m,n}(z')|$$
$$\leq|s_{m,n}(z)-s_{m,n}(z')|\leq\varphi(|z-z'|).$$

That is, each $F_{m,n}(x,y,z)$ is Rakotch contraction with respect to the third variable. So, each $w_{m,n}(x,y,z)$ is chosen so that the function $F_{m,n}(x,y,z)$ is a Rakotch contraction with respect to the third variable.

Connection 10

In view of a φ-contraction, the connections between the coefficients of variable z are obtained as follows:

1. In the affine FIS (cf. [22]), for all $t>0$,

$$\varphi(t):=\max_{m=1,2,\ldots,M}\max_{n=1,2,\ldots,N}|d_{m,n}|t,$$

where $|d_{m,n}|<1$ for all $m=1,2,\ldots,M$ and $n=1,2,\ldots,N$.

2. In the FIS with vertical scaling factors as function (cf. [21]), for all $t>0$,

$$\varphi(t):=\max_{m=1,2,\ldots,M}\max_{n=1,2,\ldots,N}\max_{x\in I}|d_{m,n}(x)|t,$$

where $d_{m,n}(x)$ is Lipschitz function defined on I satisfying $\sup_{x\in I}|d_{m,n}(x)|<1$ for all $m=1,2,\ldots,M$ and $n=1,2,\ldots,N$.

Connection 11

The continuity of bivariable FIFs differ from the continuity of univariable FIFs.

1. The graphs of linear univariable FIFs are always continuous curves.

2. There are bivariable discontinuous functions that interpolate the given data; (see for instance [23], p. 630, 631).

3. Theorem 2.3 ensures that attractors of constructed IFSs are graphs of some bounded functions which interpolate the given data, but these graphs (i.e., the graphs of bivariable FIFs) are not always continuous surfaces. Some continuity conditions of bivariable FIFs are given explicitly by Corollary 2.1 and Corollary 2.2.

Connection 12

The key difficulty in constructing fractal interpolation surfaces (or volumes) involves ensuring continuity. Another important element necessary in modelling complicated surfaces of this type is the existence of the contractivity, or vertical scaling, factors.

1. In order to ensure continuity of a fractal interpolation surface, in [22], the interpolation points on the boundary was assumed collinear, whereas in [21], vertical scaling factors as (continuous) 'contraction functions' were used.

2. A new bivariable fractal interpolation function by using the Matkowski fixed point theorem and the Rakotch contraction is presented in [18]. In order to ensure the continuity of nonlinear FIS, the coplanarity of all the interpolation points on the boundaries instead of collinearity of interpolation points on the boundary was assumed in [18], whereas in [12], vertical scaling factors as (continuous) 'contraction functions' were used.

Connection 13

1. In Theorem 2.2, we can see that

$$a_n = \frac{x_n - x_{n-1}}{x_N - x_0}, \quad b_n = \frac{x_N x_{n-1} - x_0 x_n}{x_N - x_0}$$

$$c_n = \frac{y_n - y_{n-1}}{x_N - x_0} - \frac{d_n(x_N)s_n(y_N) - d_n(x_0)s_n(y_0)}{x_N - x_0},$$

$$f_n = \frac{x_N y_{n-1} - x_0 y_n}{x_N - x_0} - \frac{x_N d_n(x_0)s_n(y_0) - x_0 d_n(x_N)s_n(y_N)}{x_N - x_0}.$$

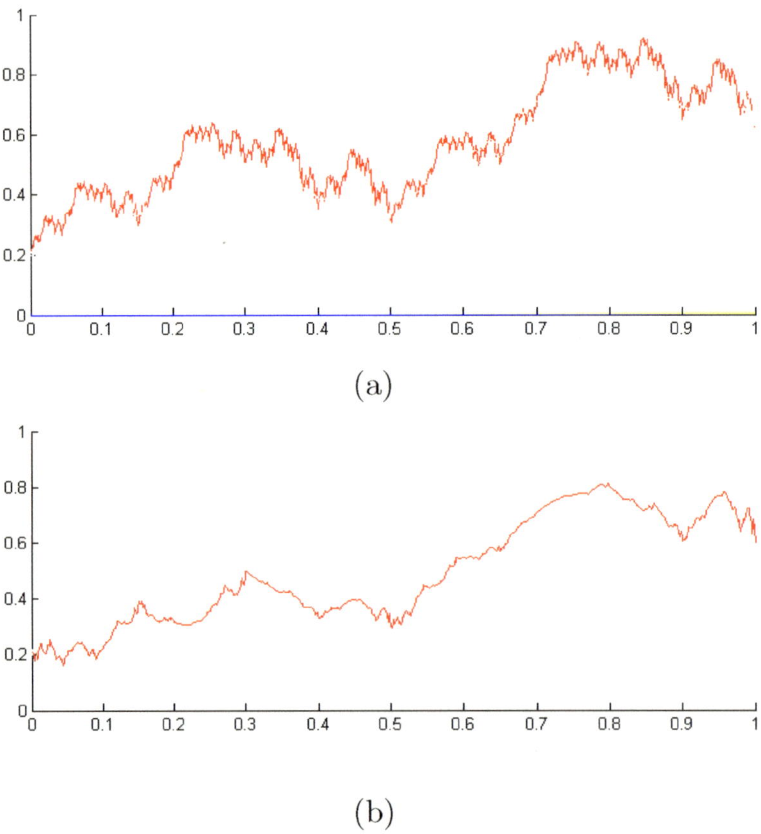

Figure 1.
The graph of a fractal interpolation function (a) that is associated with Banach contractions, (b) that is not necessarily associated with Banach contractions.

2. In Corollary 2.1, we can see that

$$a_m = \frac{x_m - x_{m-1}}{x_M - x_0}, \quad b_m = \frac{x_M x_{m-1} - x_0 x_m}{x_M - x_0},$$

$$c_n = \frac{y_n - y_{n-1}}{y_N - y_0}, \quad d_n = \frac{y_N y_{n-1} - y_0 y_n}{y_N - y_0},$$

$$g_{m,n} = \frac{(z_{m,n} - z_{m-1,n}) - (z_{m,n-1} - z_{m-1,n-1})}{(x_M - x_0)(y_N - y_0)},$$

$$e_{m,n} = \frac{y_N(z_{m,n-1} - z_{m-1,n-1}) - y_0(z_{m,n} - z_{m-1,n})}{(x_M - x_0)(y_N - y_0)},$$

$$f_{m,n} = \frac{x_M(z_{m-1,n} - z_{m-1,n-1}) - x_0(z_{m,n} - z_{m,n-1})}{(x_M - x_0)(y_N - y_0)},$$

$$h_{m,n} = \frac{x_0 y_0 z_{m,n} - x_0 y_N z_{m,n-1} - x_M y_0 z_{m-1,n} + x_M y_N z_{m-1,n-1}}{(x_M - x_0)(y_N - y_0)} - s_{m,n}(z_{M,N}).$$

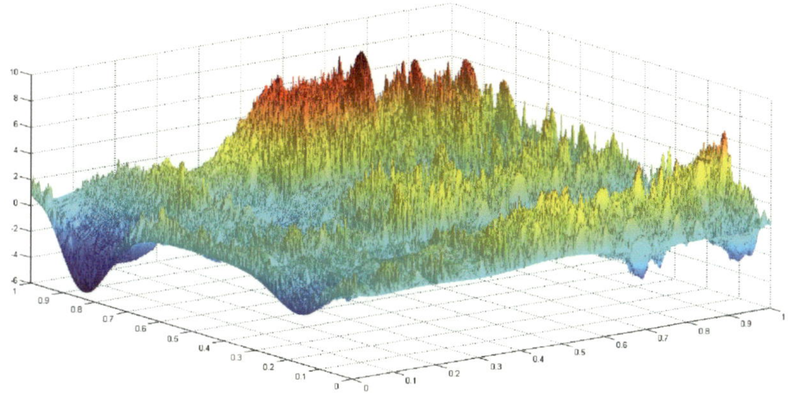

(a)

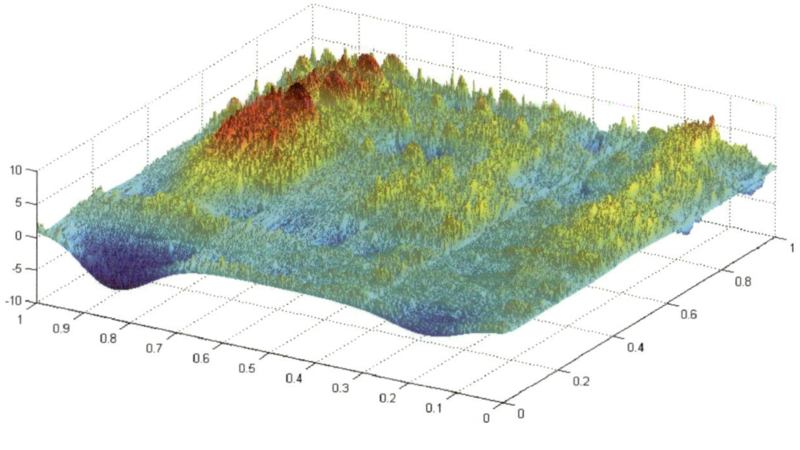

(b)

Figure 2.
A fractal interpolation surface (a) that is associated with Banach contractions, (b) that is not necessarily associated with Banach contractions.

3. In Corollary 2.2, we can see that (compare with above coefficients)

$$g_{m,n} = \frac{(z_{m,n} - z_{m-1,n}) - (z_{m,n-1} - z_{m-1,n-1})}{(x_M - x_0)(y_N - y_0)},$$

$$e_{m,n} = \frac{y_N(z_{m,n-1} - z_{m-1,n-1}) - y_0(z_{m,n} - z_{m-1,n})}{(x_M - x_0)(y_N - y_0)},$$

$$f_{m,n} = \frac{x_M(z_{m-1,n} - z_{m-1,n-1}) - x_0(z_{m,n} - z_{m,n-1})}{(x_M - x_0)(y_N - y_0)},$$

$$h_{m,n} = \frac{x_0 y_0 z_{m,n} - x_0 y_N z_{m,n-1} - x_M y_0 z_{m-1,n} + x_M y_N z_{m-1,n-1}}{(x_M - x_0)(y_N - y_0)}.$$

Figures 1(a) and **2(a)** are associated with Banach contractions, whereas **Figures 1(b)** and **2(b)** are not necessarily associated with Banach contractions.

4. Conclusions and further work

We reviewed nonlinear fractal interpolation functions by using the Geraghty fixed point theorem instead of the Banach fixed point theorem (or the Rakotch fixed point theorem) since Banach contraction (or Rakotch contraction) is a special case of Geraghty contraction. Theorems 2.1, 2.2 and 2.3 ensure that attractors of constructed nonlinear iterated function systems are graphs of some continuous functions which interpolate the given data. In particular, Examples 1, 2 and 3 show that our results remain still true under essentially weaker conditions on the maps of iterated function systems. The methods presented here can be directly extended to piecewise fractal interpolation functions that are based on recurrent IFS. A premise for future work is to extend these methods to hidden-variable fractal interpolation surfaces as well as to identify the parameters of such surfaces.

Author details

SongIl Ri[1] and Vasileios Drakopoulos[2*]

1 Department of Mathematics, University of Science, Pyongyang, D.P.R. of Korea

2 Department of Computer Science and Biomedical Informatics, University of Thessaly, Lamia, Greece

*Address all correspondence to: vdrakop@uth.gr

References

[1] Barnsley MF. Fractal functions and interpolation. Constructive Approximation. 1986;**2**:303-329

[2] Barnsley MF. Fractals Everywhere. 3rd ed. New York: Dover Publications, Inc.; 2012

[3] Chand AKB, Kapoor GP. Generalized cubic spline fractal interpolation functions. SIAM Journal on Numerical Analysis. 2006;**44**(2):655-676

[4] Viswanathan P, Chand AKB, Navascués MA. Fractal perturbation preserving fundamental shapes: Bounds on the scale factors. Journal of Mathematical Analysis and Applications. 2014;**419**:804-817

[5] Vijender N. Bernstein fractal trigonometric approximation. Acta Applicandae Mathematicae. 2018;**159**: 11-27

[6] Barnsley MF, Elton JH, Hardin DP, Massopust P. Hidden variable fractal interpolation functions. SIAM Journal on Mathematical Analysis. 1989;**20**: 1218-1242

[7] Chand AKB, Kapoor GP. Hidden variable bivariate fractal interpolation surfaces. Fractals. 2003;**11**(3):277-288

[8] Chand AKB, Kapoor GP. Stability of affine coalescence hidden variable fractal interpolation functions. Nonlinear Analysis. 2008;**68**:3757-3770

[9] Barnsley MF, Massopust PR. Bilinear fractal interpolation and box dimension. Journal of Approximation Theory. 2015; **192**:362-378

[10] Massopust PR. Fractal Functions, Fractal Surfaces and Wavelets. San Diego: Academic Press; 1994

[11] Drakopoulos V, Manousopoulos P. On non-tensor product bivariate fractal interpolation surfaces on rectangular grids. Mathematics. 2020;**8**(4):525. DOI: 10.3390/math8040525

[12] Ri S. New types of fractal interpolation surfaces. Chaos, Solitons and Fractals. 2019;**119**:291-297

[13] Rhoades BE. A comparison of various definitions of contractive mappings. Transactions of the American Mathematical Society. 1977;**226**:257-290

[14] Strobin F, Swaczyna J. On a certain generalisation of the iterated function system. Bulletin of the Australian Mathematical Society. 2013;**87**(1):37-54

[15] Xu S, Cheng S, Zhou Z. Reich's iterated function systems and well-posedness via fixed point theory. Fixed Point Theory and Applications. 2015;**2015** (1):71. DOI: 10.1186/s13663-015-0320-7

[16] Rakotch E. A note on contractive mappings. Proceedings of the American Mathematical Society. 1962;**13**:459-465

[17] Ri S. A new nonlinear fractal interpolation function. Fractals. 2017; **25**(6):1750063 (12 pages)

[18] Ri S. A new nonlinear bivariate fractal interpolation function. Fractals. 2018;**26**(4):1850054 (14 pages)

[19] Jachymski J, Jóźwik I. Nonlinear contractive conditions: A comparison and related problems. Banach Center Publications. 2007;**77**:123-146

[20] Wang H-Y, Jia-Shan Y. Fractal interpolation functions with variable parameters and their analytical properties. Journal of Approximation Theory. 2013;**175**:1-18

[21] Feng Z, Feng Y, Yuan Z. Fractal interpolation surfaces with function vertical scaling factors. Applied Mathematics Letters. 2012;**25**(11): 1896-1900

[22] Dalla L. Bivariate fractal interpolation functions on grids. Fractals. 2002;**10**(1):53-58

[23] Xie H, Sun H. The study on bivariate fractal interpolation functions and creation of fractal interpolated surfaces. Fractals. 1997;**5**(4):625-634